# Kubernetes Anti-Patterns

Overcome common pitfalls to achieve optimal deployments and a flawless Kubernetes ecosystem

**Govardhana Miriyala Kannaiah**

# Kubernetes Anti-Patterns

**Associate Group Product Manager**: Preet Ahuja

**Publishing Product Manager**: Vidhi Vashisth

**Book Project Manager**: Ashwin Kharwa

**Senior Editor**: Divya Vijayan

**Technical Editor**: Irfa Ansari

**Copy Editor**: Safis Editing

**Proofreader**: Mudita S

**Indexer**: Tejal Daruwale Soni

**Production Designer**: Shankar Kalbhor

**DevRel Marketing Coordinator**: Rohan Dobhal

First published: June 2024

Production reference: 1300524

Published by Packt Publishing Ltd.

Grosvenor House

11 St Paul's Square

Birmingham

B3 1RB, UK

ISBN 978-1-83546-068-9

www.packtpub.com

*To my twin baby boys, Jaswanth and Jayanth, for showing me how talent and creativity evolve.*
*To my family for their love, support, and inspiration.*

*– Govardhana M K*

# Contributors

## About the author

**Govardhana Miriyala Kannaiah**, founder of NeuVeu Software Private Limited, is a seasoned Kubernetes and cloud architect with a rich 16-year tenure in cloud engineering, DevOps, and technical consulting. He has spearheaded digital transformation projects through innovative solutions, assisting clients across the globe from the US to Europe and Asia with harnessing the benefits of Kubernetes and cloud technologies.

Residing in Chittoor, Andhra Pradesh, India, he holds a Bachelor of Technology degree and is a passionate industry advocate, contributing to forums and sharing knowledge on LinkedIn. He is grateful for the support of his family and the professional community that fuels his continuous pursuit of excellence.

# About the reviewers

**Pradeep Chintale** is an accomplished lead cloud solutions architect at SEI Investment Company in Pennsylvania, USA. With over 18 years of experience in the field, he specializes in system analysis, infrastructure automation, and cloud/DevOps engineering. His expertise spans across critical technology stacks including cloud, DevOps, Kubernetes, cybersecurity, and AI/MLOps. Pradeep is recognized for his exceptional ability to develop secure, reliable, and robust platforms for both private and public clouds, driving significant advancements in cloud architecture and application deployment.

Throughout his career, Pradeep has demonstrated a profound commitment to enhancing cloud solutions by fostering innovative approaches and leveraging cutting-edge technologies. At SEI Investment Company, he leads the design and implementation of Kubernetes-based container orchestration systems to ensure scalability, security, and high availability of applications. His role involves translating complex business needs into technical solutions, establishing enhanced CI/CD pipelines, integrating security checks, and managing infrastructure configurations through version-controlled methodologies. His strategic insights and technical acumen have been pivotal in streamlining operations and enhancing productivity across various departments.

Pradeep is an active member of the technical community, contributing to several professional groups including IEEE and the Linux Foundation. He is also a respected author and thought leader, having penned numerous publications and served as a speaker at several industry conferences. His contributions to the field are recognized through various awards and commendations, affirming his status as a top professional in cloud technologies and DevOps practices. As he continues to push the boundaries of cloud engineering, Pradeep remains dedicated to mentoring the next generation of engineers and shaping the future of cloud solutions.


**Viachaslau Matsukevich** has over a decade of experience in DevOps, cloud solutions architecture, and infrastructure engineering. He is certified by Microsoft, Google, and the Linux Foundation as a Solutions Architect Expert, Professional Cloud Architect, and **Certified Kubernetes Administrator (CKA)**.

Viachaslau writes about cloud-native technologies and Kubernetes, authors online courses on DevOps, and volunteers as a judge in tech awards and hackathons. He serves as an ambassador for the DevOps Institute and the Continuous Delivery Foundation, and is an Alibaba Cloud MVP, advocating for best practices and innovation in cloud computing and DevOps.

*I would like to express my deepest gratitude to my patient wife and loving family for their unwavering support and encouragement throughout this journey. Their endless patience and understanding have been a cornerstone of my success, providing me with the strength and inspiration needed to pursue my goals. Thank you for standing by my side every step of the way.*

# Table of Contents

# Part 1: Understanding Kubernetes Anti-Patterns

1

2

3

# Causes and Consequences                    39

# Part 2: Implementing Best Practices

4

5

# 8

# Proactive Assessment and Prevention                              185

9

# Bringing It All Together     209

# Index     225

# Other Books You May Enjoy     234

# Preface

Hello there! Kubernetes is a powerful platform for managing containerized applications across a cluster of machines. It is a critical tool for modern software deployment, allowing for high availability, scaling, and efficient application management. However, navigating Kubernetes can be challenging, especially when common pitfalls—known as anti-patterns—threaten the stability, efficiency, and security of applications. This book, *Kubernetes Anti-Patterns: Avert Potential Pitfalls Through a Practitioner's Lens*, is dedicated to uncovering these pitfalls, understanding their implications, and learning how to avoid or mitigate them effectively.

There are several areas where Kubernetes practitioners often encounter difficulties:

- Identifying and addressing Kubernetes anti-patterns that can degrade system performance or lead to unmanageable configurations

- Implementing best practices in Kubernetes deployments to ensure scalability, security, and maintainability

- Continuously improving Kubernetes environments to adapt to evolving needs and technologies

While numerous resources cover Kubernetes' basics and advanced features, there's a gap in literature specifically focused on anti-patterns. This book aims to fill that gap by providing a comprehensive guide on identifying anti-patterns, understanding their impacts, and adopting best practices to avert these issues. Based on my experience as a Kubernetes and cloud architect, combined with insights from leading Kubernetes professionals across various industries, this book offers practical advice and strategies derived from real-world scenarios.

As Kubernetes continues to evolve, the complexity of managing containerized applications grows. This book is an essential resource for Kubernetes practitioners looking to enhance their skills and avoid common mistakes. Whether you're a DevOps engineer, system administrator, IT manager, or software developer, you'll find valuable insights to improve your Kubernetes deployments.

## Who this book is for

- Kubernetes practitioners seeking to deepen their understanding of common pitfalls and how to avoid them

- IT professionals responsible for managing and scaling containerized applications

- Developers and system architects looking to implement best practices in Kubernetes deployments

# What this book covers

*Chapter 1, Introduction to Kubernetes Anti-Patterns*, lays the foundations by defining Kubernetes anti-patterns, their significance, and their impact on operations, guiding readers through the initial understanding necessary to navigate the complexities of Kubernetes.

*Chapter 2, Recognizing Common Kubernetes Anti-Patterns*, delves into the identification of prevalent anti-patterns, offering readers the tools to spot and understand the implications of these patterns in real-world Kubernetes environments.

*Chapter 3, Causes and Consequences*, explores the root causes of common anti-patterns and their broader effects on the Kubernetes ecosystem, equipping readers with the knowledge to address these issues at their source.

*Chapter 4, Practical Solutions and Best Practices*, provides actionable strategies and established best practices to effectively address and mitigate Kubernetes anti-patterns, ensuring optimized and resilient deployments.

*Chapter 5, Real-World Case Studies*, presents in-depth analyses of real-world scenarios where Kubernetes anti-patterns were successfully identified and resolved, illustrating the application of best practices and solutions in diverse environments.

*Chapter 6, Performance Optimization Techniques*, focuses on techniques to enhance the performance, efficiency, and scalability of Kubernetes deployments, offering insights into maximizing the potential of Kubernetes in various operational contexts.

*Chapter 7, Embracing Continuous Improvement in Kubernetes*, discusses the importance of an iterative, continuous improvement approach to Kubernetes deployment and management, highlighting strategies to adapt to the evolving Kubernetes ecosystem for sustained operational excellence.

*Chapter 8, Proactive Assessment and Prevention*, emphasizes the development of a proactive mindset for anticipating and mitigating potential pitfalls in Kubernetes environments, detailing assessment strategies and preventative measures.

*Chapter 9, Bringing It All Together*, concludes the book by synthesizing key insights, strategies, and lessons learned throughout the chapters, guiding readers on how to apply this knowledge to foster stable, efficient, and secure Kubernetes environments and encouraging a culture of continuous improvement.

# To get the most out of this book

You will need to have an understanding of the basics of Kubernetes environments, container orchestration, and the principles of DevOps practices to effectively utilize the insights from this book.

| Software/hardware covered in the book | Operating system requirements |
|---|---|
| Kubernetes | Windows, macOS, or Linux |
| Docker | Windows, macOS, or Linux |
| Cloud providers (AWS, Azure, and Google Cloud) | |
| Continuous integration tools | Windows, macOS, or Linux |
| Monitoring tools (Prometheus, Grafana, etc.) | Windows, macOS, or Linux |

## Conventions used

There are a number of text conventions used throughout this book.

`Code in text`: Indicates code words in text, database table names, folder names, filenames, file extensions, pathnames, dummy URLs, user input, and Twitter handles. Here is an example: "Mount the downloaded `WebStorm-10*.dmg` disk image file as another disk in your system.

A block of code is set as follows:

```
Dockerfile
# Stage 1: Build the application
FROM node:16 as builder
```

Any command-line input or output is written as follows:

```
kubectl apply -f nginx-ingress.yaml
```

**Bold**: Indicates a new term, an important word, or words that you see onscreen. For instance, words in menus or dialog boxes appear in **bold**. Here is an example: "Select **System info** from the **Administration** panel."

> **Tips or important notes**
> Appear like this.

## Get in touch

Feedback from our readers is always welcome.

**General feedback**: If you have questions about any aspect of this book, email us at customercare@packtpub.com and mention the book title in the subject of your message.

**Errata**: Although we have taken every care to ensure the accuracy of our content, mistakes do happen. If you have found a mistake in this book, we would be grateful if you would report this to us. Please visit www.packtpub.com/support/errata and fill in the form.

**Piracy**: If you come across any illegal copies of our works in any form on the internet, we would be grateful if you would provide us with the location address or website name. Please contact us at copyright@packt.com with a link to the material.

**If you are interested in becoming an author**: If there is a topic that you have expertise in and you are interested in either writing or contributing to a book, please visit authors.packtpub.com.

## Share Your Thoughts

Once you've read *Kubernetes Anti-Patterns*, we'd love to hear your thoughts! Scan the QR code below to go straight to the Amazon review page for this book and share your feedback.

https://packt.link/r/1835460682

Your review is important to us and the tech community and will help us make sure we're delivering excellent quality content.

# Download a free PDF copy of this book

Thanks for purchasing this book!

Do you like to read on the go but are unable to carry your print books everywhere?

Is your eBook purchase not compatible with the device of your choice?

Don't worry, now with every Packt book you get a DRM-free PDF version of that book at no cost.

Read anywhere, any place, on any device. Search, copy, and paste code from your favorite technical books directly into your application.

The perks don't stop there, you can get exclusive access to discounts, newsletters, and great free content in your inbox daily

Follow these simple steps to get the benefits:

1.  Scan the QR code or visit the link below

https://packt.link/free-ebook/9781835460689

2.  Submit your proof of purchase
3.  That's it! We'll send your free PDF and other benefits to your email directly

# Part 1: Understanding Kubernetes Anti-Patterns

In this part, you will gain a comprehensive understanding of Kubernetes anti-patterns, including their origins, impacts, and how to identify them in practical scenarios. This part covers essential strategies for recognizing and addressing common anti-patterns to optimize Kubernetes deployments effectively.

This part contains the following chapters:

- *Chapter 1, Introduction to Kubernetes Anti-Patterns*
- *Chapter 2, Recognizing Common Kubernetes Anti-Patterns*
- *Chapter 3, Causes and Consequences*

# 1

# Introduction to Kubernetes Anti-Patterns

In the ever-evolving landscape of container orchestration and cloud-native technologies, Kubernetes stands as a shining beacon, offering unparalleled power and flexibility in managing containerized applications. Yet, this power comes with a price – the complexity of Kubernetes can sometimes lead even the most experienced practitioners down treacherous paths, resulting in suboptimal deployments, operational headaches, and, ultimately, compromised application performance.

Kubernetes, often referred to as K8s, is an intricate ecosystem of interconnected components, each with its own set of best practices, configurations, and potential pitfalls. While this intricacy empowers organizations to build highly resilient and scalable applications, it also creates opportunities for missteps. Kubernetes environments are not immune to the allure of shortcuts, temporary fixes, or misconfigurations, leading to issues that fall into the category of anti-patterns.

Before we delve into the specifics of Kubernetes anti-patterns, it's essential to understand the roadmap that lies ahead. The subsequent chapters will take you through a series of anti-patterns, shedding light on the challenges they present and offering guidance on how to navigate past them.

Enter the concept of Kubernetes anti-patterns. We'll embark on a journey to dissect and demystify the notion of anti-patterns in the context of Kubernetes. Just as cartographers study treacherous terrain to chart safer paths, Kubernetes practitioners must learn to recognize and mitigate anti-patterns to navigate their Kubernetes landscapes successfully.

Let's explore the diverse and sometimes daunting terrain of Kubernetes anti-patterns together. It's time to make your Kubernetes journey smoother, more efficient, and ultimately more successful.

We'll cover the following topics in this chapter:

- Understanding Kubernetes anti-patterns
- The significance of identifying anti-patterns
- The impact across the Kubernetes ecosystem

# Understanding Kubernetes anti-patterns

At its core, an anti-pattern is a practice, configuration, or strategy that might initially seem like a good idea but, in the long run, undermines the overall health and performance of your system. Anti-patterns are the sly adversaries that threaten the integrity of your Kubernetes deployment. They present a paradoxical challenge – they are intuitively appealing or expedient in the short term, making them attractive even to seasoned practitioners. This is where the danger lies as they act as hidden traps, luring both the uninitiated and the experienced alike.

In the context of Kubernetes, these anti-patterns are not mere nuisances; they can lead to a cascade of issues, from inefficiencies in resource utilization to a significant drop in application performance, and, ultimately, operational chaos. This is why understanding and identifying Kubernetes anti-patterns is of paramount importance for organizations looking to make the most of their Kubernetes investments.

But why are Kubernetes anti-patterns crucial to understand? They serve as cautionary tales and lessons learned, preventing you from making costly mistakes that can jeopardize your Kubernetes deployment's success. By recognizing and addressing these anti-patterns, you safeguard the efficiency, resilience, and adherence to best practices within your Kubernetes ecosystem. In essence, they become signposts that guide you away from the pitfalls and toward a smoother, more secure Kubernetes journey.

## The deceptive allure of anti-patterns

Kubernetes anti-patterns are tricky. They appear as easy solutions for managing Kubernetes, a system used for handling lots of applications and containers. At first glance, these anti-patterns look like they'll make your work easier and faster. But, the truth is, they're not as helpful as they seem.

These anti-patterns are sneaky. Like a chameleon blending into its surroundings, they fit into your Kubernetes setup without looking wrong. At first, you might not notice anything is off. They seem like smart, efficient ways to handle tasks in Kubernetes. But, over time, their true nature becomes clear.

The real issue with these anti-patterns is the problems they cause down the line. What starts as a seemingly clever shortcut can lead to bigger issues in how your Kubernetes system runs. You might find your applications not working as smoothly or needing more resources than you planned for.

To steer clear of these pitfalls, it's crucial to have a solid grasp of Kubernetes. Knowing the best practices helps you identify and avoid these misleading shortcuts. It's also important to regularly check on your Kubernetes setup. This helps you catch any problems early on, preventing them from growing into larger issues.

## Types and forms of Kubernetes anti-patterns

In the multifaceted world of Kubernetes, anti-patterns take on various forms, each concealing its unique pitfalls. Understanding these diverse manifestations is crucial as it allows you to appreciate the full spectrum of challenges that might lurk within your Kubernetes deployment.

Let's delve into the different types and forms of Kubernetes anti-patterns:

- **Suboptimal resource utilization**: This is one of the more prevalent anti-patterns where resources are either over-provisioned or underutilized, leading to inefficiency and wasted infrastructure costs

- **Security oversights**: These anti-patterns neglect security best practices, making your Kubernetes cluster vulnerable to threats and breaches

- **Overcomplicated architectures**: Anti-patterns can manifest as complex, convoluted architectures that defy the principle of simplicity, causing maintenance headaches and impeding scalability

- **Scaling strategies gone awry**: In the quest for scalability, anti-patterns may encourage improper scaling strategies that overburden your cluster or result in underperforming applications

- **Configuration confusion**: Some anti-patterns emerge from misconfigurations that compromise the stability and predictability of your Kubernetes deployments

## A call to vigilance

The need for vigilance against anti-patterns is not a passing concern but a constant and recurring theme. It's a call to action – a reminder that success in managing Kubernetes is not a one-time achievement but an ongoing journey that demands continuous attention and awareness.

The essential nature of vigilance arises from the chameleon-like, ever-adapting quality of anti-patterns. They are not static adversaries; they morph and evolve, mirroring the changes and challenges that Kubernetes deployments face. As your containerized ecosystem grows and transforms to meet new demands, so do the anti-patterns that threaten it. This means that even the most seasoned practitioners can fall victim to these concealed traps if they let their guard down.

Moreover, Kubernetes itself is an open, extensible platform that allows for a wide range of configurations and customizations. While this flexibility is one of its strengths, it also creates a fertile ground for anti-patterns to take root. Without vigilance, you might inadvertently introduce anti-patterns while configuring or expanding your Kubernetes cluster, even with the best of intentions.

Anti-patterns might also emerge over time as you add new workloads, scale your applications, or make updates to your system. This means that staying vigilant isn't only about being cautious during the initial setup; it's an ongoing commitment to monitor, assess, and adapt as your Kubernetes environment evolves.

Being vigilant involves several key practices:

- **Regular audits and reviews**: Periodically review your Kubernetes configurations, deployment strategies, and best practices to ensure they align with your evolving requirements.

- **Stay informed**: Keep up with Kubernetes updates, security advisories, and community best practices. Knowledge is your greatest asset in the battle against anti-patterns.

- **Collaboration**: Engage with the Kubernetes community and peers to share experiences and insights. Collaboration can uncover potential anti-patterns and their solutions.

- **Automated checks**: Implement automated checks, monitoring, and alerting systems to promptly identify and respond to deviations from best practices.

By embracing a culture of vigilance, you create a proactive and resilient Kubernetes environment. You turn anti-patterns from lurking threats into opportunities for continuous improvement. This ongoing commitment to vigilance is your safeguard against the deceptive allure and chameleon-like nature of anti-patterns. It ensures that your Kubernetes deployment remains not only efficient and secure but also adaptive and aligned with evolving best practices, ultimately leading to long-term success in the world of container orchestration.

# The significance of identifying anti-patterns

As we established in the previous section, Kubernetes anti-patterns represent enticing yet treacherous pitfalls that can undermine the health and performance of your deployments. In this section, we'll explore the critical importance of recognizing and addressing these anti-patterns.

## Guardians of stability

Within the complex realm of Kubernetes, a platform renowned for its unparalleled scalability and resilience, we find ourselves navigating a landscape that demands both our awe and our caution. Kubernetes stands as the paragon of container orchestration, the de facto choice for those seeking to manage containerized applications with finesse. It is a powerhouse that promises automated scaling, fault tolerance, and resource optimization, allowing organizations to meet the relentless demands of modern, cloud-native applications.

Yet, this immense power brings forth an inherent responsibility. As the wise adage goes, *With great power comes great responsibility*. Kubernetes is not a tool to be wielded carelessly; it requires meticulous planning, deep expertise, and strict adherence to best practices to unlock its true potential.

In the heart of this thriving ecosystem, we encounter anti-patterns – subtle adversaries that challenge the very stability and predictability that Kubernetes vows to deliver. These anti-patterns, often concealed beneath the guise of practical solutions, can erode the foundation of your Kubernetes system.

They represent practices, configurations, or strategies that, though initially appearing appealing or expedient, quietly undermine the strength and resilience of Kubernetes. Instead of clarity, they introduce complexity; in place of efficient resource utilization, they lead to wastage, and instead of smooth operations, they pave the way for chaos.

The role of anti-patterns is as fascinating as it is treacherous. It serves as a poignant reminder that while Kubernetes offers unmatched capabilities, it is equally important to recognize the potential pitfalls that could compromise your system's stability and performance. Anti-patterns, in their enigmatic role, compel us to approach Kubernetes with the utmost vigilance and expertise. They emphasize that understanding the nuances and intricacies of this dynamic ecosystem is paramount.

As we journey deeper into our exploration of Kubernetes anti-patterns, let this understanding lay the foundation for our quest. Let it serve as a beacon, illuminating the path to harnessing the full power of Kubernetes while maintaining the stability and predictability that underpin its renowned reputation.

## The butterfly effect in Kubernetes

In our exploration of Kubernetes anti-patterns, it's crucial to grasp the interconnected nature of this container orchestration ecosystem. Kubernetes is not just a collection of isolated components; it's an ever-evolving, dynamic, and interdependent web of technologies, configurations, and decisions.

Consider, for a moment, the butterfly effect in chaos theory. It posits that the flapping of a butterfly's wings in one part of the world could set a series of events in motion that eventually culminate in a hurricane on the other side of the planet. In the Kubernetes realm, this idea finds a compelling parallel.

Kubernetes is not a monolithic entity; it's a complicated ecosystem where each component, each configuration choice, and each operational decision creates ripples. A seemingly minor misstep, much like the butterfly's wing flap, can have far-reaching consequences.

Picture this: you make a small misconfiguration in a pod's resource limits, believing it to be inconsequential. Over time, this seemingly minor oversight causes the pod to consume more resources than expected, leading to resource shortages. As a result, applications in your cluster start to slow down, causing user dissatisfaction. This performance dip sets off alarms, leading to resource scaling and load distribution changes.

Now, imagine that in response to this, another resource allocation is misconfigured, exacerbating the issue and causing further performance degradation. It's like a domino effect, with each erroneous choice leading to more significant problems, akin to the butterfly's wing flap setting off a chain reaction of atmospheric disturbances.

The heart of this analogy is the interconnectedness and interdependence of Kubernetes components and practices. One misstep can, indeed, lead to a cascade of issues that reverberate across your entire infrastructure. It highlights the sensitivity of the Kubernetes environment and the profound impact that individual decisions can have.

Understanding the butterfly effect in Kubernetes is not just about being aware of the potential pitfalls but recognizing the need for meticulous planning, continuous monitoring, and the deployment of best practices. It is a reminder that, in this dynamic ecosystem, every action and decision has the potential to create a chain reaction. As we delve deeper into the world of Kubernetes anti-patterns, let this awareness guide our approach to ensuring the stability, efficiency, and resilience of our deployments.

## Efficiency and resource optimization

Efficiency and resource optimization are the underpinning pillars of a successful and cost-effective deployment. It's not just a matter of buzzwords; it's about making the most of every resource at your disposal. Kubernetes is, after all, a finely tuned machine where every component must perform harmoniously to deliver optimal results.

Imagine your Kubernetes cluster as a well-tuned vehicle's engine, where resources, such as CPU, memory, storage, and more, are allocated judiciously to support applications without waste or deficiency. When resources are not optimally managed, the implications ripple through your entire ecosystem.

Resource allocation is a delicate dance, and striking the right balance is paramount. Overprovisioning is a common pitfall, where more resources than needed are allocated. This might seem like a safety net for peak loads, but it often results in unnecessary operational costs. On the flip side, underutilization occurs when resources are insufficiently allocated, leading to inefficiency as applications struggle with limited resources.

Now, consider Kubernetes anti-patterns – those stealthy foes that often emerge as misconfigurations, scaling errors, or inefficient resource utilization. They have a knack for disrupting this resource equilibrium. By not addressing these anti-patterns, it's like trying to save water with the tap wide open – wasteful and detrimental to your budget and operational capabilities.

Identifying and mitigating these anti-patterns are pivotal steps to optimizing resources within your Kubernetes environment. This isn't just about cutting costs; it's about ensuring that your applications operate at peak performance without resource-related bottlenecks or overages.

Resource optimization unlocks several essential benefits:

- **Reduced operational costs**: Efficient resource utilization translates into lowered infrastructure expenses, extending to operational costs. You're not paying for underused resources, and applications perform optimally with allocated resources.

- **Enhanced scalability**: Properly optimized resources empower efficient application scaling. With judicious resource allocation, scaling becomes precise and responsive, addressing application demands without wasting capacity.

- **Better quality of service**: Resource-optimized applications run smoothly and responsively. Users enjoy consistent performance, and you're more likely to meet or surpass **service-level agreements (SLAs)**.

- **Sustainability**: By avoiding overprovisioning, you contribute to environmental sustainability by reducing power consumption and associated carbon emissions.

Efficiency and resource optimization are not mere ideals; they are practical necessities in the realm of Kubernetes. The journey to identify and address Kubernetes anti-patterns is a commitment to resource efficiency, cost-effectiveness, and the realization that by optimizing your resource allocation, you can indeed achieve the enviable position of doing more with less.

## Reliability and performance

The reliability of your applications and the performance they deliver are the yardsticks by which the success of your Kubernetes ecosystem is often measured.

Consider Kubernetes as a stage where your applications perform intricate ballets, and user expectations are set high. The stage must be both reliable and agile to meet these expectations. However, the presence of Kubernetes anti-patterns challenges this equilibrium, often casting a shadow over the performance and reliability of your applications.

Reliability is an assurance that your applications will be available and function as expected, consistently and without interruption. In the dynamic realm of modern applications, where uptime is non-negotiable, reliability is paramount.

Kubernetes, with its promises of resilience and high availability, sets the stage for such reliability. But when anti-patterns are allowed to accumulate, they become the lurking adversaries that undermine these promises. A cluster marred by anti-patterns is like a stage that trembles – a stage that cannot guarantee the consistent performance your users expect.

Performance, the other half of the equation, is about more than just delivering results; it's about delivering results with speed and efficiency. Users today demand instantaneous responses, and applications must meet these expectations to remain competitive.

Anti-patterns within your Kubernetes ecosystem are not merely nuisances; they are the hidden traps that can lead to unscheduled downtime, sluggish response times, and, ultimately, disgruntled users. They can manifest as inefficient resource allocation, poor scaling decisions, or even security oversights that impact the performance of your applications.

Recognizing and addressing these anti-patterns is not just good practice; it is central to preserving the high availability and responsiveness that Kubernetes promises. It ensures that the stage upon which your applications perform remains steady and that the applications themselves dazzle with peak performance.

Moreover, it's essential to recognize that anti-patterns don't exist in isolation. Much like the butterfly effect in chaos theory, one anti-pattern can set off a chain reaction of issues. For instance, a misconfiguration that leads to resource bottlenecks can strain the entire cluster, impacting every application running on it. The ripple effect can lead to cascading issues, affecting not just one application but many.

In this interdependent ecosystem, where each component and configuration influences the entire system, even minor issues can escalate, causing widespread disruption. This emphasizes the need for a vigilant approach in identifying and mitigating anti-patterns to maintain reliability and performance.

## Security and compliance

Security and compliance are not just watchwords; they are the sentinels guarding the fortress of your digital assets. In a world where cyber threats loom large and regulatory frameworks are continually evolving, security and compliance are not merely important; they are imperative.

Kubernetes, with its promise of robust security features and best practices, serves as a stronghold for applications and data. It offers a fortified environment that can withstand cyber assaults, but this security can be jeopardized by the stealthy intruders known as Kubernetes anti-patterns.

Security is the assurance that your digital fortress is protected from unauthorized access, data breaches, and cyber-attacks. Kubernetes, with its built-in security mechanisms and the ability to implement best practices, provides a secure sanctuary for your applications and data. It's a place where vulnerabilities are actively monitored and threats are deflected.

However, when anti-patterns infiltrate your Kubernetes environment, they serve as subtle chinks in the armor. They might introduce misconfigurations, overlook security best practices, or create vulnerabilities that can be exploited. These can lead to security breaches, data leaks, and unmitigated cyber threats, endangering the integrity of your applications and the trust of your users.

Beyond just security, regulatory compliance is an essential concern. Industries are subject to various legal frameworks, standards, and best practices. These rules are in place to protect consumers, secure data, and ensure the ethical use of technology. Failing to adhere to these regulations can result in severe consequences, including legal penalties and damage to reputation.

Kubernetes, when configured and operated according to industry standards and best practices, provides a robust platform that can help you meet compliance requirements. It's a powerful ally in your journey of navigating the complex labyrinth of legal frameworks.

Kubernetes anti-patterns are not just operational nuisances; they are hidden threats to security and compliance. They might introduce misconfigurations that expose your cluster to vulnerabilities, opening the door to malicious actors. They can create unnecessary complexities that make it challenging to maintain and enforce compliance, inadvertently putting your organization at risk.

Recognizing and addressing these anti-patterns becomes more than a good practice; it's an imperative. It involves safeguarding your fortress, protecting sensitive data, and assuring your organization operates within the boundaries of the law.

## Maintainability and scalability

Maintainability and scalability are the unsung heroes of your operations. They are the unseen gears that keep your Kubernetes deployment running smoothly and allow it to evolve as your organization's needs change. The presence or absence of anti-patterns has a profound impact on these crucial aspects of your Kubernetes ecosystem.

Maintainability is the quality that ensures your Kubernetes deployment can be managed and kept in a healthy, operational state with ease. It's about being able to apply updates, make changes, and troubleshoot issues without significant disruption. In a fast-paced world where applications constantly evolve, maintainability is the unsung hero that keeps your environment agile.

Kubernetes, with its declarative configuration and self-healing capabilities, offers a strong foundation for maintainability. It simplifies operations and allows for efficient resource management. But when anti-patterns infiltrate your environment, they introduce complexities, misconfigurations, and operational inefficiencies, making the environment more challenging to maintain.

Scalability, on the other hand, is the engine that powers your organization's growth. It's about being able to handle increased workloads, more users, and expanding resource needs. Kubernetes is renowned for its scalability, offering the ability to handle growing applications seamlessly.

But anti-patterns, when left unaddressed, can throw a wrench into the scalability machinery. These stealthy culprits can lead to overutilized resources, inefficient scaling decisions, or even security vulnerabilities that hinder your ability to scale as needed. This not only impacts your growth but can lead to application performance bottlenecks.

Addressing anti-patterns proactively is not just about fixing problems; it's about preventing them. By identifying and mitigating anti-patterns, you ensure that your Kubernetes environment remains maintainable and scalable. This proactive approach simplifies operational tasks, reduces the risk of issues arising, and paves the way for your applications to grow seamlessly.

The synergy between maintainability and scalability cannot be overstated. A maintainable environment is a prerequisite for efficient scaling. When you have a system that is easy to manage and operate, you can confidently scale up your applications to meet increasing demand.

## Cost control and resource allocation

Cost control and resource allocation are the financial underpinnings of your operation. They are the levers you can pull to strike a balance between efficient resource utilization and budgetary prudence. The presence or absence of anti-patterns has a profound impact on the financial aspects of your Kubernetes ecosystem.

Resource allocation is akin to managing a budget. It's about ensuring that you allocate just the right number of resources to your applications, no more and no less. Overallocation results in needless spending, while underallocation leads to operational inefficiency. It's a tightrope walk that requires precision.

Kubernetes, with its ability to optimize resource allocation, provides a strong foundation for cost control. It allows you to allocate resources based on the actual needs of your applications. However, when anti-patterns infiltrate your environment, they disrupt this delicate balance. Misconfigurations, inefficient scaling, or security oversights can lead to resource wastage or bottlenecks, impacting both your budget and service quality.

Cost control, in turn, is about prudently managing your financial resources. It's a fundamental concern for any organization. Wasteful spending erodes profitability and can impact the overall financial health of your organization.

Anti-patterns, when not addressed, become the silent culprits that drive up operational costs. Overallocation of resources leads to higher infrastructure expenses, while underallocation can result in operational inefficiency, indirectly affecting your bottom line.

Identifying and mitigating anti-patterns not only helps maintain a delicate balance between resource allocation and cost control but also has a significant impact on your organization's financial health. This prudent approach ensures that you allocate resources accurately, reducing unnecessary infrastructure costs while maintaining the quality of service. The savings can be substantial, directly impacting your organization's bottom line.

## A competitive advantage

In the fast-paced, ever-competitive landscape of modern technology and business, gaining an edge can be the difference between success and obscurity. In the context of Kubernetes and its associated anti-patterns, organizations that excel in recognizing and managing these hidden challenges are well-positioned to secure a potent competitive advantage.

Kubernetes, with its promise of scalability, resilience, and efficient resource management, offers organizations a solid foundation for deploying applications and services. It's a technology that can empower businesses to meet the demands of today's digitally-driven world, where customers expect seamless experiences and high-quality services.

However, Kubernetes is a double-edged sword. While it empowers, it also demands expertise and vigilance. Without a keen understanding of the potential pitfalls and challenges it poses, organizations may find themselves grappling with operational inefficiencies, soaring costs, and reduced service quality. This is where Kubernetes anti-patterns come into play.

Organizations that invest in recognizing and managing Kubernetes anti-patterns gain a substantial competitive edge. Here's how:

- **Faster innovation**: By effectively addressing anti-patterns, you streamline your operations and free up resources. This agility allows your teams to focus on innovation rather than firefighting. You can roll out new features, services, and updates more rapidly, staying ahead of the curve in a rapidly changing market.

- **Higher service quality**: Addressing anti-patterns directly impacts the reliability, performance, and security of your services. With these critical aspects well-maintained, you can provide your customers with higher-quality experiences. Happy, satisfied customers are more likely to remain loyal and recommend your services to others.

- **Cost-effective operations**: Efficient resource allocation and reduced operational costs go hand in hand with recognizing and mitigating anti-patterns. This financial prudence can be a game-changer for your organization since it enhances profitability and allows you to invest in strategic initiatives.

- **Superior customer experiences**: High-performing, reliable services not only retain existing customers but also attract new ones. A superior customer experience can be a potent differentiator in a crowded market, often trumping lower prices or flashy marketing.

In the competitive landscape of modern business, excellence in recognizing and managing Kubernetes anti-patterns is more than just a matter of best practices; it's about seizing a competitive edge. Organizations that master these challenges are poised to innovate faster, offer superior services, and deliver exemplary customer experiences, all while maintaining a cost-effective operational model.

In this section, we explored the significance of identifying anti-patterns. It's not merely a matter of best practices; it's about safeguarding your Kubernetes environment, your applications, and your organization's competitive position. As we journey through this chapter and beyond, we will equip you with the tools and knowledge needed to not only identify these anti-patterns but to also strategically eliminate them. This, in turn, will empower you to unlock the full potential of Kubernetes while ensuring the stability, efficiency, and security of your deployments.

# The impact across the Kubernetes ecosystem

The influence of Kubernetes anti-patterns reverberates far beyond the confines of individual deployments. In this section, we'll explore how these stealthy adversaries can cast their shadows across the entire Kubernetes ecosystem.

Anti-patterns are rarely contained within a single Kubernetes cluster. It functions as a dynamic ecosystem where clusters and components interact to deliver applications and services. For instance, a misconfigured application in one cluster can create a surge in traffic to other clusters, leading to resource shortages and potentially affecting service quality for multiple users.

## Performance degradation

In the sprawling ecosystem of Kubernetes, efficiently allocating and sharing resources is pivotal. Clusters often act as nodes in a network, sharing a common pool of vital resources, including network bandwidth, storage, and computational power. These shared resources ensure the smooth operation of applications and services, facilitating seamless interactions between clusters and components.

Anti-patterns often introduce inefficiencies in resource allocation. For instance, overconsumption due to misconfigurations or inefficient resource usage can lead to a significant imbalance within a cluster. As a result, this cluster starts consuming a disproportionate number of shared resources.

This skewed resource allocation doesn't remain contained within the problematic cluster. It propagates across the Kubernetes ecosystem, impacting interconnected clusters and their applications. As the overburdened cluster strains shared resources, it causes resource scarcity in other clusters. The result is a cascading effect, whereby performance degradation becomes pervasive.

Performance degradation isn't confined to a single cluster; it becomes a shared burden. Applications and services running in interconnected clusters experience slowdowns, service interruptions, and increased latency. The entire Kubernetes ecosystem begins to show signs of inefficiency, affecting user experiences and the ability to meet operational demands.

Understanding the repercussions of such resource scarcity underscores the critical importance of proactively identifying and mitigating anti-patterns. It's the key to maintaining a harmonious balance and ensuring optimal performance across the Kubernetes ecosystem.

## Maintenance complexity

Maintaining a Kubernetes ecosystem is a multifaceted endeavor that requires precision, proactive monitoring, and effective troubleshooting. However, the presence of anti-patterns significantly increases maintenance complexity, turning the operational maze into an intricate labyrinth of challenges. In this section, we'll delve into how anti-patterns amplify the complexity of maintenance and explore the dimensions of this challenge.

Anti-patterns often lead to operational issues, creating a need for constant troubleshooting. These issues can range from misconfigurations to inefficiencies in resource management, necessitating time and effort to diagnose and rectify. Frequent troubleshooting diverts valuable resources away from core operational tasks and can lead to an environment of perpetual firefighting.

Effective maintenance relies on proactive monitoring to detect issues before they impact services. However, the presence of anti-patterns increases the need for continuous monitoring. Their subtle, hidden nature means that vigilance is paramount, requiring resources dedicated to ongoing, vigilant scrutiny.

Maintenance also involves keeping the Kubernetes ecosystem up to date with the latest patches and updates. Anti-patterns can complicate the update process. They might introduce dependencies or configurations that clash with new versions, leading to potential downtime, compatibility issues, and operational disruptions.

Maintenance complexity often necessitates a broader skill set. Anti-patterns introduce unique challenges that require in-depth knowledge and expertise to navigate. Operators and administrators may need to learn new strategies and workarounds to address these challenges effectively.

Addressing these challenges proactively becomes essential to simplifying maintenance, reducing resource drains, and ensuring a more stable and secure Kubernetes environment.

# Developer productivity

In the world of Kubernetes, developer productivity isn't just a matter of efficiency; it's the linchpin of innovation and the engine that drives the rapid deployment of applications and services. However, the presence of anti-patterns can be a significant obstacle to developer productivity in several crucial ways.

Developers are at the forefront of building and maintaining applications in a Kubernetes ecosystem. When anti-patterns lead to operational issues, it's often the development teams that get pulled into the troubleshooting process. This diversion of developer focus from core development tasks to firefighting can hinder productivity.

Efficient testing is a critical component of the software development life cycle, ensuring the quality and reliability of applications. Anti-patterns can lead to unstable environments, delaying the testing and deployment of applications. Developers need a stable platform for rigorous testing, and anti-pattern-induced instability can disrupt the iterative development process.

Kubernetes already has a learning curve, and anti-patterns can exacerbate this. Developers may need to navigate the intricacies that are introduced by these anti-patterns, learning new strategies and workarounds to manage the challenges they pose. This learning curve can slow down development and adaptation to best practices.

To create a developer-friendly Kubernetes environment, organizations must actively address anti-patterns to empower their development teams and streamline the process of building and deploying applications. This is essential for staying competitive in a fast-paced, technology-driven world.

# Interoperability challenges

The Kubernetes ecosystem is just one part of a broader technological landscape that includes various tools, services, and platforms. Achieving seamless interoperability with these external components is a key objective in modern IT operations. However, anti-patterns can throw a wrench into this critical aspect of your Kubernetes ecosystem, posing a range of challenges that impact the overall functionality of your applications and operational efficiency:

- **Third-party integrations**: Kubernetes is designed to be extensible, allowing you to integrate with third-party tools and services seamlessly. However, anti-patterns can disrupt these integrations. For instance, misconfigured or overly complex **custom resource definitions** (**CRDs**) may hinder the ability to interact with external services and technologies. This can limit your organization's ability to harness the full potential of Kubernetes for orchestration and automation.

- **Data exchange**: Effective data exchange is crucial for modern applications. Anti-patterns that affect data flow and exchange mechanisms can hinder your Kubernetes ecosystem's ability to communicate with external databases, message queues, or data analytics platforms. This can impede real-time data processing, analysis, and reporting, potentially undermining your application's functionality.

- **Monitoring and observability**: Kubernetes and its associated anti-patterns can impact your observability and monitoring stack. Misconfigured or incompatible monitoring agents may result in incomplete or inconsistent data collection, making it difficult to diagnose and troubleshoot issues that arise across your integrated systems.

- **Orchestration and workflow tools**: Kubernetes anti-patterns can also impact your orchestration and workflow tools. For example, complex or suboptimal CI/CD pipelines may hinder the automation of deployment and scaling processes, affecting the agility of your applications and slowing down the development cycle.

- **Service mesh integration**: Service meshes such as Istio or Linkerd are essential for managing communication between microservices. Anti-patterns, such as inadequate configuration or incomplete deployment of service mesh components, can disrupt traffic routing, security policies, and observability features, impacting the reliability and security of your applications.

- **Container registry access**: Efficient access to container images in a container registry is crucial for smooth application deployment. Anti-patterns that affect registry authentication, image pull policies, or image storage can cause delays and bottlenecks during image retrieval, impacting your application's startup time.

- **Authentication and authorization**: Security anti-patterns within your Kubernetes environment can also affect authentication and authorization mechanisms when integrating with external systems. These issues can result in inconsistent user access controls and data security concerns when interacting with other services.

Addressing interoperability challenges caused by anti-patterns requires a holistic approach. It involves aligning your Kubernetes practices with industry standards, ensuring proper integration testing, and actively monitoring external system interactions to identify and rectify potential issues.

## Long-term technical debt

The failure to address anti-patterns can lead to the accumulation of a subtle yet insidious burden known as long-term technical debt. Technical debt refers to the cumulative cost of delaying necessary work on your systems. It's the interest that accrues as you choose shortcuts, bypass best practices, and opt for expedient solutions, all of which, over time, can transform your Kubernetes ecosystem into a complex, difficult-to-maintain landscape.

Here's how long-term technical debt manifests in the context of Kubernetes:

- **Complexity accumulation**: Anti-patterns often introduce complexity into your Kubernetes setup. Whether it's convoluted configurations, workarounds, or temporary fixes, these complexities can accumulate. Over time, this leads to a tangled web of interdependencies, making it challenging to understand, modify, or scale your environment.

- **Innovation hurdles**: Long-term technical debt hinders your ability to innovate. Your team ends up spending a significant portion of their time managing existing systems and addressing issues arising from anti-patterns. This leaves little room for exploring new technologies, implementing best practices, or developing cutting-edge features, which are essential for staying competitive in today's rapidly changing landscape.

- **Competitive disadvantage**: Organizations that allow long-term technical debt to accumulate can find themselves at a competitive disadvantage. They struggle to adapt to new industry trends, meet changing customer demands, and deliver features and updates quickly. In contrast, competitors who have managed their Kubernetes environments effectively can seize market opportunities more readily.

- **Diminished reliability**: Over time, anti-patterns can undermine the reliability and availability of your applications. Unaddressed issues, such as poor scaling strategies, configuration errors, and lack of redundancy, can result in service disruptions, causing frustration for users and damaging your reputation.

- **Increased costs**: Technical debt is not just a hidden cost; it can become a significant financial burden. As your Kubernetes ecosystem becomes more complex, operational costs can rise, from increased labor hours being required to maintain the environment to potential expenses related to service disruptions or security breaches.

Addressing long-term technical debt requires a proactive and strategic approach. This includes periodically conducting comprehensive audits of your Kubernetes environment, prioritizing the resolution of anti-patterns, and investing in continuous improvement.

## Summary

In this inaugural chapter, we explored the essence of Kubernetes anti-patterns from various angles and delved into their deceptive nature, adaptability, and the diverse forms they can take. We also emphasized the call to remain vigilant against them, recognizing the constant need for awareness and scrutiny in the world of Kubernetes.

Understanding these anti-patterns is merely the first step in this exploration. We unveiled how the allure of anti-patterns, even to seasoned practitioners, can lead to shortcuts and solutions that ultimately undermine the system's health and performance. We drew parallels between anti-patterns and chameleons, highlighting how they adapt and blend into their environment until they strike unexpectedly.

Moreover, we delved into the types and forms that anti-patterns can take, encompassing everything from configuration blunders to inefficient scaling practices. By acknowledging their diversity, we are better equipped to recognize and address these pitfalls as they arise.

The significance of identifying anti-patterns becomes evident as we discuss their role as the guardians of stability, their potential to set off a butterfly effect that ripples throughout your Kubernetes environment, and their far-reaching implications on efficiency, reliability, security, maintainability, and cost control. These factors underscore why vigilance against anti-patterns is not merely a best practice but a critical necessity.

To cap off this chapter, we examined how anti-patterns impact the entire Kubernetes ecosystem. Whether it's through degraded performance, elevated maintenance complexity, decreased developer productivity, interoperability challenges, or the accrual of long-term technical debt, these adversarial practices ripple through the Kubernetes landscape, affecting your applications, services, and the very foundation of your Kubernetes deployment.

With this foundational understanding of Kubernetes anti-patterns, we lay the groundwork for a deeper exploration of specific anti-patterns in the subsequent chapters. Armed with the knowledge of their characteristics, significance, and impact, we are ready to navigate the complex terrain of Kubernetes anti-patterns and embark on a mission to recognize, address, and ultimately overcome these hidden traps.

In the next chapter, we'll explore common pitfalls in Kubernetes environments, including over-reliance on pod-level resources and inefficient networking configurations. Additionally, real-world scenarios, tools for identification, and the consequences of these anti-patterns will be discussed, stressing the need for proactive mitigation.

# 2

# Recognizing Common Kubernetes Anti-Patterns

Recognizing and understanding common anti-patterns within your Kubernetes infrastructure is akin to illuminating potential disruptions that can compromise the stability and functionality of your system. This chapter acts as a comprehensive guide, unveiling prevalent stumbling blocks within Kubernetes setups and delving deep into their origins, defining characteristics, and the profound disruptive impact they exert on the smooth operation of Kubernetes environments.

This undertaking involves a meticulous examination of core issues that threaten the seamless and optimal performance of Kubernetes setups. It's about more than just identifying problems; it's an opportunity to gain a deeper understanding of the intricate complexities and nuances within these architectures. This exploration enables a proactive approach, empowering individuals to not only recognize these issues but also to troubleshoot and resolve them effectively.

Understanding these anti-patterns offers more than a list of what to avoid; it provides a roadmap toward improved practices. By acknowledging what doesn't work optimally, individuals and teams can craft strategies and implementations that align with proven successful methodologies. This fosters an environment of continuous improvement, nurturing innovation within Kubernetes architectures.

The ultimate goal is to equip system administrators, DevOps teams, platform engineering professionals, and Kubernetes practitioners with the knowledge and foresight to preemptively detect, effectively manage, and prevent these detrimental patterns. This proactive approach aims to fortify and elevate the reliability, resilience, and overall efficiency of Kubernetes ecosystems, creating a more stable and optimized operational environment.

We'll cover the following topics in this chapter:

- Ten common anti-patterns in Kubernetes
- Identifying anti-patterns in real-world scenarios
- Real consequences of anti-patterns

# Ten common anti-patterns in Kubernetes

Within Kubernetes environments, a set of prevalent anti-patterns can profoundly impact the efficiency and reliability of deployments. Recognizing these 10 common anti-patterns is a critical step for professionals seeking to proactively manage and enhance the performance and stability of their Kubernetes infrastructures.

## 1. Over-reliance on pod-level resources

Kubernetes heavily relies on the effective allocation and management of resources at the pod level to enhance application performance. However, an excessive dependency on these resources can lead to numerous adverse patterns that significantly influence the system's overall health and stability.

One notable issue arising from over-reliance on pod-level resources is the lack of effective resource utilization patterns. Overemphasizing resource allocation within individual pods without considering inter-pod communication and resource sharing may result in inefficient use of available resources. This lack of holistic resource utilization can lead to underutilization and hinder the overall performance efficiency of the entire system.

Furthermore, strict adherence to fixed resource assignments within pods can create rigidity. When resources are allocated in a rigid, unalterable manner within pods, it may restrict the system's ability to adapt to varying workloads or demands. This inflexibility could limit the system's responsiveness and resilience, impacting its overall performance in dynamic environments.

Inadequate coordination of resource distribution among pods might lead to bottlenecks or resource imbalances. Over-relying on individual pod-level resource management without considering how resources are distributed across multiple pods might result in uneven resource utilization, leading to potential bottlenecks and inefficiencies across the cluster.

Moreover, a lack of standardized resource-sharing mechanisms between pods might create disparities. Overemphasis on individual pod resources without standardized sharing protocols can result in resource monopolization, causing disparities in resource availability and hindering the system's overall performance.

## 2. Misusing or overusing ConfigMaps and Secrets

ConfigMaps and Secrets are essential components in Kubernetes, facilitating the management of configuration data and sensitive information. However, improper use or overuse of these resources can introduce significant challenges, particularly concerning security and operational complexities within the Kubernetes environment.

ConfigMaps primarily store configuration data in key-value pairs, allowing decoupling of configuration from container images. This separation enables easier configuration updates without changing the core application. On the other hand, Secrets are designed specifically for storing sensitive information, such as passwords, tokens, and encryption keys, in a more secure manner.

Misusing ConfigMaps often involves excessive reliance on them for storing large chunks of data that could be more suitably managed elsewhere. While ConfigMaps are excellent for configuration settings, they are not optimized for storing large volumes of data. Inefficient usage leads to increased pod startup times and, in extreme cases, can even cause issues such as API server timeouts.

Overusing ConfigMaps could lead to a cluttered and disorganized system, making it harder to maintain and manage configurations effectively. When multiple ConfigMaps are created for each individual configuration change, it can become challenging to track, maintain, and understand the overall system configuration.

Similarly, mishandling Secrets involves storing non-sensitive data in Secrets, which defeats their primary purpose of securing sensitive information. Such misuse can lead to confusion and potential security risks, especially during debugging or code reviews.

Furthermore, using Secrets with inadequate security measures, such as storing plaintext passwords or sensitive information without encryption, poses a considerable risk. If an unauthorized entity gains access to these Secrets, it could compromise the entire system's security.

## 3. Monolithic containerization

The concept of containerization in Kubernetes is centered on breaking down applications into smaller, more manageable components. However, the anti-pattern of monolithic containerization arises when entire monolithic applications are encapsulated within containers, leading to various inefficiencies and challenges.

A typical monolithic application consists of multiple modules or services that can function independently. However, in the context of containerization, these monolithic applications are placed within a single container, contradicting the fundamental philosophy of microservices and containerization principles.

The drawbacks of this approach include limitations in scalability and resource inefficiencies. Monolithic containerization restricts the scalability potential that microservices architecture offers. Scaling the entire monolith becomes less efficient compared to the granularity achievable with individual microservices.

Resource inefficiencies arise when deploying a monolithic container. Resources are allocated for the entire application, even if certain modules require significantly fewer resources. This leads to inefficient use of resources and restricts the ability to optimize resource allocation.

Additionally, deployment complexity increases with monolithic containerization. Updates or modifications to a monolithic container necessitate deploying the entire application, even when changes might affect specific modules. This elongates deployment times and introduces the risk of errors in the process.

Furthermore, managing dependency conflicts becomes a challenge. Monolithic containers might face issues with dependency conflicts, especially when different modules within the monolith require varying versions of libraries or software components. This can lead to complexities in managing dependencies and compatibility issues.

## 4. Lack of resource limits and quotas

Efficient resource management is pivotal for maintaining system stability and preventing potential issues.

When resource limits and quotas are not adequately defined, several issues can emerge. Firstly, without defined resource limits, certain pods or containers might consume excessive resources, leading to resource contention within the cluster. This contention can cause performance degradation and affect other applications sharing the same resources.

The absence of resource limits can lead to unpredictable behavior within the system. Pods with high resource demands might starve others, resulting in unexpected downtime or failures, making the system less reliable.

Moreover, the absence of enforced quotas makes capacity planning and resource management challenging. Predicting future resource needs or preventing potential overloads within the cluster becomes challenging, hindering the scalability and growth of the Kubernetes environment.

Security risks also loom large when resources are left unrestricted. Unchecked resource consumption might lead to potential security vulnerabilities and abuse. An attacker, either intentionally or unintentionally, might leverage excessive resources, causing a **denial of service** (**DoS**) for other legitimate applications.

## 5. Ignoring pod health probes

Ensuring the health and reliability of applications is critical for system stability. The anti-pattern of ignoring pod health probes presents several challenges that can compromise system resilience and reliability.

Pod health probes, such as readiness and liveness probes, play a vital role in determining the health status of pods within the cluster. Ignoring or improperly configuring these probes can result in various issues.

The readiness probe is responsible for determining when a pod is prepared to handle traffic. If this probe is disregarded or misconfigured, it can allow traffic to be directed to a pod before it's fully ready. This premature traffic influx can lead to service disruptions or errors, especially when the pod is not in a stable state.

On the other hand, the liveness probe checks whether a pod is running as expected. Neglecting this probe or setting it up incorrectly can result in malfunctioning pods continuing to receive traffic even when they are unresponsive or have failed.

A consequence of ignoring pod health probes is difficulty in identifying failing pods effectively. This can result in degraded service quality and reliability as the Kubernetes system continues to route traffic to pods that may not be functioning correctly.

## 6. Bloated container images

Container images are the fundamental building blocks for applications. The anti-pattern of bloated container images occurs when these images contain unnecessary or excessive components, leading to various inefficiencies and challenges.

A bloated container image often contains redundant or oversized elements that inflate its size without providing proportional benefits. Such inefficiencies in container images can lead to several issues.

Firstly, bloated container images result in increased network latency and longer deployment times due to their larger size. Pulling and deploying these images can consume more bandwidth and storage space, leading to slower image transfers and longer startup times.

Moreover, larger image sizes often impact resource utilization. They consume more memory and storage space in the Kubernetes cluster, leading to inefficiencies in resource allocation and potentially hindering the performance of the overall system.

Security risks also increase with bloated container images. Larger images not only introduce potential vulnerabilities but also expand the attack surface, as more components within the image might present security risks.

## 7. Overutilization of Persistent Volumes

**Persistent Volumes** (**PVs**) provide a way for applications to access durable storage resources. However, the anti-pattern of overutilizing PVs occurs when these resources are excessively or inefficiently employed, leading to several challenges within the system.

One common issue stemming from the overutilization of PVs is the inadequate allocation or inefficient usage of storage resources. When PVs are overused, they might be allocated beyond actual application needs, leading to wasted resources and increased costs.

Additionally, overutilization might result in storage contention, where multiple applications or pods are competing for the same PV. This contention can cause performance degradation and impact the reliability of applications relying on those resources.

Improper monitoring and lack of efficient resource utilization policies can exacerbate the problem. When PVs are overutilized and not efficiently managed, it becomes challenging to predict future storage requirements or prevent potential overloads within the Kubernetes cluster.

## 8. Unnecessary resource sharing among microservices

In the context of Kubernetes and microservices architecture, an anti-pattern emerges when microservices unnecessarily share resources. While the modularity and autonomy of microservices typically involve distinct and independent functionality, unnecessary resource sharing between these services can lead to various inefficiencies and complexities within the system.

Resource sharing among microservices can include the unnecessary sharing of databases, caches, or other resources. Although inter-service communication and collaboration are essential, sharing resources that are not vital for service functionality can lead to several challenges.

Firstly, unnecessary resource sharing can result in increased dependencies between microservices. When services share resources beyond their core functionality, any changes or modifications to these shared resources might impact multiple microservices, leading to complexities in managing these dependencies.

Moreover, it can hinder the scalability and flexibility of microservices. When services share resources, the scalability of one service might be impacted by the load or behavior of another service, reducing the independence and autonomy that microservices aim to provide.

Security risks also increase with unnecessary resource sharing. Exposing resources to multiple services might amplify vulnerabilities, creating a larger attack surface that can compromise the security of the entire system.

## 9. Inefficient or over-complicated networking configurations

Networking configurations play a crucial role in ensuring the proper communication and connectivity between various components. However, the anti-pattern of inefficient or over-complicated networking configurations introduces challenges that can affect system performance, scalability, and maintenance.

This anti-pattern often arises due to overly complex network setups or inefficient use of networking resources. When networking configurations are needlessly intricate, they can lead to several issues within the Kubernetes environment.

Firstly, complex networking setups might result in difficulties in managing, maintaining, and troubleshooting the network. Overly convoluted configurations can make it challenging to understand the network topology, diagnose issues, and implement changes effectively.

Moreover, inefficient networking configurations can lead to suboptimal performance. Misconfigurations or over-complicated setups might result in increased latencies or bottlenecks, hindering the overall performance and responsiveness of applications within the cluster.

In addition, over-complicated networking can lead to increased operational overhead. Unnecessarily intricate configurations might require more time and effort for regular maintenance and can become a barrier to scaling the network effectively.

## 10. Overlooking Horizontal Pod Autoscaling opportunities

**Horizontal Pod Autoscaling (HPA)** is a powerful feature in Kubernetes that dynamically adjusts the number of running instances of a given application, based on the observed CPU utilization or other configurable metrics. However, the anti-pattern of overlooking HPA opportunities occurs when users fail to leverage this feature effectively, missing out on the potential benefits it offers.

Ignoring or underutilizing HPA can lead to various challenges and missed optimization opportunities within the Kubernetes ecosystem.

Firstly, overlooking HPA means missing the chance to automatically scale applications in response to varying workloads. Failure to implement HPA can result in underutilization of resources during periods of low demand or overloading of the system during high-demand situations.

Moreover, not utilizing HPA can lead to inefficient resource allocation. When the number of pod instances remains static and doesn't scale based on actual needs, it can lead to over-provisioning of resources, wasting computing power, and incurring unnecessary costs.

Additionally, ignoring HPA opportunities can impact system performance and reliability. When an application doesn't automatically adjust its resources according to varying workloads, it may result in slower response times or even service interruptions during peak loads.

Having explored 10 common anti-patterns in Kubernetes, we now have a clearer understanding of potential pitfalls that can disrupt our Kubernetes environments. These patterns, ranging from over-reliance on pod-level resources to overlooking HPA, highlight the key areas where vigilance is essential. As we move forward, we will shift our focus to *Identifying anti-patterns in real-world scenarios*. This next section is designed to take our theoretical knowledge into practical scenarios, showing how these anti-patterns can appear in real-world Kubernetes deployments. We'll learn how to spot these patterns in action, understand their practical consequences, and discover strategies to avoid and resolve them, ensuring a more stable and efficient Kubernetes environment.

# Identifying anti-patterns in real-world scenarios

Understanding the signs, causes, and implications of these real-world instances is pivotal in proactively addressing, mitigating, and preventing these anti-patterns in Kubernetes infrastructures. This section aims to provide valuable insights into recognizing and effectively managing these anti-patterns for enhanced system performance, scalability, and reliability.

## Monitoring and metrics for resource overutilization

The primary objective of monitoring and metrics for resource overutilization is to track and analyze resource usage. This involves monitoring key metrics such as CPU utilization, memory usage, and network throughput to identify potential overutilization issues within the Kubernetes cluster.

Implementing effective monitoring tools enables continuous tracking of resource metrics. Prometheus, Grafana, and Kubernetes-native tools such as `kube-state-metrics` are commonly used to collect and visualize resource utilization data. These tools help in identifying spikes or consistently high usage patterns that signify potential overutilization.

Setting appropriate thresholds is essential to trigger alerts when resource utilization exceeds defined limits. Alerts notify administrators when resource usage reaches critical levels, enabling timely intervention to rectify the overutilization.

Using monitoring and metrics, administrators can identify pods, services, or nodes causing resource overutilization. It allows for the implementation of mitigation strategies such as workload redistribution, resource tuning, or optimization of application code to resolve identified overutilization issues.

Furthermore, historical data from monitoring and metrics provide insights into trends, facilitating capacity planning and proactive resource allocation adjustments to prevent future instances of overutilization.

By leveraging effective monitoring and metrics tools, Kubernetes users can promptly detect, analyze, and address resource overutilization issues, ensuring the efficient utilization of resources and enhancing the overall stability and performance of their Kubernetes environments.

## Audit and compliance tools for Secrets and configurations

Audit and compliance tools for Secrets and configurations play a crucial role in maintaining a secure environment. The primary objective is to enforce and verify adherence to security policies and compliance requirements.

These tools enable continuous auditing and monitoring of Secrets, configuration files, and access permissions. They track changes, access attempts, and configurations to identify potential security gaps or unauthorized modifications. Audit logs provide a historical record of actions taken, aiding in forensic analysis and identifying security breaches or compliance violations.

Utilizing tools such as **Open Policy Agent** (**OPA**), Kubernetes Secrets, and ConfigMap controllers, administrators can define and enforce policies for Secret and configuration management. Policies might include access controls, encryption standards, and validation requirements to ensure compliance with security standards and industry regulations.

Implementing automated checks and periodic audits ensures that Secrets and configurations meet defined policies and compliance standards. Continuous monitoring and regular audits help detect deviations from established guidelines and immediately alert administrators to take corrective actions.

Furthermore, integrating these audit and compliance tools with **identity and access management** (**IAM**) systems helps enforce **role-based access control** (**RBAC**) and restrict unauthorized access to Secrets and configurations.

The effective implementation of audit and compliance tools for Secrets and configurations ensures a proactive approach to security, enabling administrators to maintain a secure and compliant Kubernetes environment. By identifying and rectifying potential vulnerabilities, these tools contribute to the overall robustness and trustworthiness of the system.

## Assessment strategies for containerization practices

Assessment strategies for containerization practices involve evaluating the containerization approach to ensure optimal efficiency and performance. This process assists in identifying areas of improvement and potential anti-patterns related to containerized applications.

One key element of assessing containerization practices is conducting a thorough review of container images. This includes analyzing the image size and layers and identifying unnecessary components. Tools such as Docker Slim or `dive` assist in examining image layers and identifying redundant elements that contribute to image bloat.

Assessment also involves evaluating the application architecture and its alignment with microservices principles. Assessing whether applications are appropriately decomposed into microservices helps in determining scalability, maintainability, and resource utilization efficiency.

Analyzing container orchestration settings and resource allocation is vital in ensuring optimal performance. Assessment tools such as Kubernetes-native resources and tools provided by cloud providers enable administrators to evaluate and fine-tune resource settings for better utilization.

Security and compliance assessment is another critical aspect. Evaluating security measures within containers, such as image scanning for vulnerabilities or verifying compliance with best practices, contributes to a more secure environment.

Additionally, performance assessment by conducting load testing and benchmarking helps identify potential bottlenecks and performance limitations within containerized applications.

Regularly conducting these assessments enables administrators to identify potential anti-patterns and areas for improvement in containerization practices. Implementing findings from these assessments ensures a more efficient, scalable, and secure Kubernetes environment. Assessments contribute to the continual optimization of containerization practices, aligning them with best practices and improving overall system performance.

## Visibility into resource limitation and quota management

Effective visibility into resource limitation and quota management involves comprehensive monitoring, enforcement, and governance of resource usage.

Monitoring tools such as Prometheus, Grafana, and native Kubernetes monitoring capabilities offer insights into resource consumption trends. They provide visibility into CPU, memory, storage, and network usage, enabling administrators to identify usage patterns and potential overconsumption.

Setting and enforcing resource quotas for namespaces or specific workloads is a fundamental part of resource management. Lack of quotas or inadequate limits might result in some applications consuming more resources than necessary, potentially impacting the performance of other applications.

Visibility into existing quotas and their enforcement requires robust governance practices. Utilizing Kubernetes tools such as `ResourceQuota` and `LimitRange` allows administrators to establish and enforce quotas effectively.

Implementing alerts and notifications when resource quotas are nearing their limits ensures proactive measures to prevent resource exhaustion. These alerts help administrators take corrective actions, such as scaling resources or optimizing workloads before reaching critical resource limits.

Continuous review and adjustment of quotas based on workload changes and performance requirements are essential. Regular assessments ensure that allocated resources align with the actual needs of applications running in the cluster.

A comprehensive view of resource limitation and quota management ensures a balanced allocation of resources, prevents resource contention, and maintains a stable and efficient Kubernetes environment. It provides administrators with the insights necessary to optimize resource utilization and prevent potential resource-related anti-patterns from affecting the system's stability.

## Health probe monitoring and alerting mechanisms

Effective health probe monitoring and alerting mechanisms involve constant surveillance and timely alerts for pod health states, ensuring that only healthy pods serve traffic.

The readiness and liveness probes within Kubernetes are critical for assessing the operational status of pods. Neglecting these probes or failing to configure them correctly can result in directing traffic to pods that may not be fully prepared to handle requests or are unresponsive, causing service disruptions.

Implementing health probe monitoring involves continuous checks to verify the readiness and liveness of pods. Tools such as Kubernetes events and probes, along with monitoring platforms such as Prometheus, enable administrators to continuously track pod health statuses.

Configuring alerting mechanisms is essential to respond promptly to failing or unresponsive pods. Setting up alerts that trigger notifications when a pod fails readiness or liveness checks allows for immediate investigation and remediation.

Regular testing and simulation of different scenarios ensure that health probes accurately reflect the actual state of pods. This practice helps in identifying potential issues before they affect live services.

Proactive remediation of failing pods and corrective actions such as scaling, restarting, or deploying redundant pods ensure that the service remains uninterrupted and maintains optimal performance.

Prioritizing and maintaining a robust health probe monitoring and alerting system within Kubernetes is imperative for ensuring the continuous health and stability of applications. Implementing these mechanisms aids in preventing service disruptions and upholding a reliable and resilient Kubernetes environment.

## Image optimization techniques for efficient containerization

Image optimization techniques play a critical role in managing container images to reduce their size while maintaining functionality.

Analyzing and reducing image size is fundamental to image optimization. Tools such as Docker Slim or multi-stage builds in Dockerfiles help minimize image size by removing unnecessary components, unused packages, and layers.

Implementing efficient caching mechanisms during image building reduces the need to rebuild unchanged components, speeding up the build process and reducing deployment times.

Utilizing smaller base images and shared layers aids in minimizing the overall size of images. Alpine Linux and other minimal base images provide a lightweight foundation for container images.

Regular updates and patches to images ensure security and reduce vulnerabilities. Automating the image update process ensures that images remain secure and up to date without manual intervention.

Implementing image scanning tools such as Clair or Trivy helps identify and mitigate security vulnerabilities within container images, ensuring a more secure and reliable environment.

Continuous performance testing and benchmarking of optimized images ensure that they perform optimally and do not introduce performance bottlenecks in the system.

By adopting these image optimization techniques, administrators can significantly reduce image size, resource overhead, and deployment times while improving security and performance within their Kubernetes environment. Optimized images contribute to a more efficient and robust containerization process, enhancing the system's overall efficiency and security.

## Audit tools for PV management

Audit tools for PV management involve monitoring, tracking changes, and maintaining the integrity and security of data stored in PVs.

Effective monitoring tools enable continuous tracking and assessment of PVs. Tools such as Kubernetes Volume Snapshots, `kubectl describe` commands, or specific storage vendor tools provide insights into volume states, resource consumption, and any potential issues.

Tracking changes within PVs is crucial for maintaining data integrity. Audit logs and versioning mechanisms help administrators track modifications, ensuring that changes are intentional and within compliance standards.

Regular backups and snapshots of PVs ensure data resilience and recovery. Implementing automated backups using tools such as Velero allows for the efficient restoration of data in case of volume failure or accidental data loss.

Ensuring security and compliance within PVs is essential. Regular audits ensure encryption, access controls, and compliance with security standards are maintained. Tools such as Aqua Security or Sysdig assist in assessing and maintaining security within PVs.

Establishing alerts for critical events, such as volume capacity nearing limits or unauthorized access attempts, is vital for immediate action and proactive maintenance of PVs.

By leveraging effective audit tools for PV management, administrators can maintain the integrity, security, and efficiency of persistent storage within their Kubernetes environment. Proper auditing contributes to a more resilient and secure data storage system, reducing the likelihood of data loss and potential vulnerabilities.

## Analysis of service-to-service resource sharing

The analysis of service-to-service resource sharing involves evaluating the degree of resource sharing and identifying potential anti-patterns to maintain service autonomy and optimal performance.

Examining the level of resource sharing between microservices is critical. Analyzing the extent to which services share resources, databases, caches, or components helps in understanding the dependencies and potential risks associated with over-sharing.

Identifying unnecessary resource sharing is essential. Services should share only vital resources required for communication, ensuring that unnecessary dependencies and potential performance impacts are minimized.

Assessing the impact of resource sharing on service autonomy and scalability is pivotal. Analyzing how resource sharing affects the independence and scalability of microservices helps in understanding the system's overall efficiency and potential anti-patterns.

Implementing strict controls and governance to manage and restrict unnecessary resource sharing ensures that services maintain autonomy and do not create unnecessary interdependencies that could compromise the system's reliability.

Regular reviews and audits of service-to-service resource-sharing practices help in identifying potential bottlenecks or inefficiencies arising from excessive sharing. Adjustments and optimizations based on these assessments aid in improving the system's performance and scalability.

By conducting a thorough analysis of service-to-service resource sharing, administrators can mitigate potential anti-patterns, optimize service autonomy, and enhance the overall efficiency and reliability of their microservices architecture within the Kubernetes environment. Efficient resource-sharing practices contribute to a more scalable and robust system without unnecessary interdependencies.

## Network analysis tools for identifying complex configurations

Utilizing network analysis tools for identifying complex configurations is crucial to streamline communication and troubleshoot potential issues within the network.

Analysis of network configurations involves utilizing tools such as Kubernetes-native networking features, network plugins, or specialized tools such as Wireshark to scrutinize communication pathways and potential complexities within the network.

Identifying bottlenecks or network congestion points is vital for ensuring efficient traffic flow. Analysis tools help pinpoint these areas, enabling administrators to take corrective actions to optimize network traffic and prevent potential communication issues.

Evaluating DNS resolution and service discovery mechanisms aids in ensuring smooth service communication. Complexities within these processes can lead to service disruptions or communication failures, making it essential to identify and streamline these configurations.

Assessing load-balancing configurations helps in maintaining an even distribution of traffic and preventing overload on specific components. Tools such as `kube-proxy` or service mesh tools facilitate load-balancing analysis.

Continuous monitoring and periodic audits of network configurations ensure that the network setup aligns with the system's evolving needs. Regular assessments help in identifying and rectifying potential complexities that could impact overall system performance.

By leveraging network analysis tools to identify complex configurations, administrators can streamline communication pathways, address potential bottlenecks, and optimize network settings within the Kubernetes environment. Efficiencies in network configuration contribute to enhanced system performance and reliability.

## Metrics and triggers for autoscaling opportunities

Effective implementation of autoscaling relies on defining metrics and triggers to scale resources efficiently based on workload variations.

Defining appropriate metrics, such as CPU utilization, memory consumption, or custom application-specific metrics, is fundamental for autoscaling. Tools such as HPA or custom Prometheus queries assist in setting up and monitoring these metrics.

Establishing triggers based on predefined thresholds ensures timely resource adjustments. Trigger configurations, set through HPA or custom scripts, prompt the system to scale resources up or down in response to workload changes.

Continuous monitoring of workload patterns aids in identifying potential autoscaling opportunities. Analyzing historical data and trends enables administrators to anticipate workload changes and adjust scaling parameters proactively.

Implementing predictive scaling strategies based on forecasting workload trends assists in preemptive resource adjustments, minimizing the impact of sudden workload changes.

Load testing and simulation of different workload scenarios help in fine-tuning autoscaling configurations. Verifying how the system responds to varying workloads ensures that autoscaling mechanisms are effective and reliable.

By defining accurate metrics, establishing appropriate triggers, and continuously monitoring and refining autoscaling strategies, administrators can ensure a responsive and optimally scaled Kubernetes environment. Effective autoscaling not only prevents resource underutilization but also mitigates the risk of system overload during peak demands, leading to a more efficient and cost-effective system operation.

# Real consequences of anti-patterns

In this section, we delve deeper into tangible repercussions that arise from the prevalence of Kubernetes anti-patterns. Understanding the real-life implications and consequences of these patterns is crucial in appreciating their impact on the reliability, scalability, and maintainability of Kubernetes environments.

## Operational chaos caused by configuration drift

Configuration drift in a Kubernetes environment poses a significant threat to operational stability, potentially disrupting the reliability and consistency of the entire system. It involves deviations between the actual configurations and settings of the system and the intended or desired state. In the dynamic and highly flexible realm of Kubernetes, where numerous components interact and evolve, configuration drift manifests in various forms, leading to substantial operational challenges.

The consequences of configuration drift can be severe. Inconsistencies across the cluster can result in discrepancies in application performance, potential security vulnerabilities, and difficulties in pinpointing and resolving issues. For instance, when specific settings vary across nodes or pods due to drift, it can lead to unexpected behavior or failures, making it challenging to identify the root cause of problems.

These inconsistencies can cause operational chaos, impeding smooth deployment, scaling activities, and routine operations. They may lead to downtime, performance degradation, or even security breaches, affecting the overall reliability and predictability of the Kubernetes ecosystem.

## Compliance risks and regulatory challenges

Compliance risks and regulatory challenges loom as substantial obstacles, introducing complexities that can significantly hinder operational efficiency and jeopardize system stability. The consequences of non-compliance with industry standards, data protection regulations, or internal policies are far-reaching. In failing to meet these stringent compliance standards, Kubernetes environments face an increased vulnerability to breaches, data compromise, or legal ramifications.

The inherent dynamism and fluidity of Kubernetes add layers of complexity to the challenge. The constantly evolving nature of containerized applications, coupled with the distributed, interconnected architecture of Kubernetes, amplifies the risks. The rapid deployment of microservices and containers makes it inherently challenging to maintain compliance across the entire infrastructure. The decentralized nature of these environments often leads to difficulties in enforcing consistent controls and policies, exposing vulnerabilities that could lead to non-compliance issues.

Non-compliance not only jeopardizes data security but also poses a threat to the organization's reputation and trust. Should sensitive data be compromised or regulations breached, the aftermath could be damaging, resulting in legal penalties, loss of customer trust, and substantial financial repercussions. Rectifying such breaches often requires extensive resources, time, and effort.

Addressing these risks and regulatory challenges demands a proactive approach, entailing a comprehensive understanding of the regulatory landscape and the establishment of robust governance and compliance frameworks within Kubernetes deployments. It involves continuous monitoring, stringent access controls, and consistent enforcement of security protocols to ensure compliance.

It requires a multifaceted strategy that not only focuses on adhering to regulations but also integrates security measures and policies into the very fabric of the Kubernetes infrastructure to safeguard against potential risks, thereby ensuring operational efficiency while meeting regulatory requirements.

## Lost opportunities for resource optimization

The failure to capitalize on resource optimization opportunities translates into missed potential for efficient utilization, resulting in cascading effects that impact operational efficiency and cost-effectiveness. The essence of Kubernetes lies in its ability to dynamically allocate and manage resources. However, when optimization opportunities are overlooked, inefficiencies emerge, hindering the full realization of this dynamic resource orchestration.

The consequences of overlooking resource optimization opportunities in Kubernetes are multifaceted. Inadequate resource allocation or misconfigurations lead to underutilization or over-provisioning of resources, which significantly impacts performance and scalability. Underutilization results in wasted resources, adding unnecessary operational costs and reducing overall system efficiency. On the contrary, over-provisioning not only leads to increased infrastructure expenses but can also cause performance bottlenecks and decreased system stability.

Moreover, the failure to capitalize on resource optimization opportunities impedes the ability to scale efficiently and limits the responsiveness of the Kubernetes environment to fluctuating workloads. It constrains the platform's ability to adapt swiftly to demands, hindering the organization's agility and competitiveness in the market.

Overlooked opportunities for resource optimization result in missed potential savings and diminished operational capabilities. In a highly competitive business landscape where efficiency and scalability are key differentiators, such missed opportunities can result in increased operational costs and decreased productivity.

Addressing these challenges demands a comprehensive approach involving continuous monitoring, thorough performance analysis, and robust resource management strategies. Employing automated tools for workload optimization and implementing best practices in resource allocation and utilization are pivotal.

## Service degradation and end user impact

The occurrence of service degradation not only impacts the system internally but also significantly influences end user experiences, potentially leading to severe consequences. Service degradation, when left unaddressed, causes disruptions, hindering the reliability and functionality of applications and services. As a result, end users might encounter issues such as slow response times, increased latencies, or, in severe cases, service unavailability.

The implications of service degradation are manifold, extending beyond just technical challenges. End users rely on consistent and dependable service delivery. When degradation occurs, it affects user experience, potentially leading to frustration, dissatisfaction, and, in worst-case scenarios, loss of trust in the provided services or applications.

Kubernetes, with its dynamic nature and the decentralized orchestration of containers and microservices, adds complexity to monitoring and maintaining service reliability. Service degradation might occur due to a variety of factors, including resource contention, misconfigurations, or bottlenecks in the network, among others. Addressing these challenges is complex, as pinpointing the root cause of degradation can be time-consuming in the intricate and distributed architecture of Kubernetes.

The impacts are not solely limited to end users. Service degradation can also affect the organization's reputation and financial well-being. A damaged reputation can lead to decreased customer retention and potentially hinder new customer acquisition. Financially, the repercussions might include direct revenue loss and increased support costs due to user-reported issues.

Resolving service degradation and mitigating its effects necessitates proactive and strategic measures. Implementing robust monitoring tools, ensuring adequate capacity planning, and employing automation for quick response to fluctuating workloads are crucial.

## Systemic complexity and increased maintenance efforts

Systemic complexity within a Kubernetes environment amplifies the intricacy of managing the system, creating a myriad of challenges that significantly elevate maintenance efforts. The multifaceted and interconnected nature of Kubernetes, with its diverse array of nodes, services, and pods, contributes to an environment where complexities can rapidly compound.

The sprawling nature of Kubernetes environments leads to increased maintenance overhead. As the system grows in scale and complexity, managing configurations, maintaining proper networking, and ensuring security across the entire architecture become increasingly challenging. These complexities often result in an augmented cognitive load for system administrators and operators, making routine tasks more time-consuming and error-prone.

With numerous interdependencies, identifying the root cause of problems becomes a daunting task. This intricate environment demands a deeper understanding of the interactions between various components, which, in turn, elevates the difficulty level for maintenance and problem resolution.

The increased systemic complexity within Kubernetes environments further necessitates continuous efforts in skill development and resource allocation. It requires ongoing training for personnel and additional resources for monitoring, maintenance, and troubleshooting.

Addressing these challenges involves strategic planning and the implementation of comprehensive management strategies. Embracing best practices such as consistent documentation, automated monitoring, and effective training programs can help mitigate the effects of systemic complexity.

Employing automation tools for routine tasks and ensuring a structured and organized approach to system maintenance can significantly reduce burdens associated with systemic complexities.

## Resource wastage and increased operational costs

Inefficient resource allocation and underutilization can incur substantial costs, affecting both the operational budget and overall system performance. When resources are underused or over-provisioned, the impact ripples across various aspects of the environment.

The consequences of resource wastage are multi-fold. Underutilization, where resources are not optimally utilized, results in unnecessary operational costs. Wasted resources, including unused compute capacity or storage, directly impact the bottom line, increasing operational expenses without contributing to enhanced performance or service delivery. Conversely, over-provisioning leads to an unnecessary increase in infrastructure expenses, driving up operational costs and contributing to a reduction in cost efficiency.

Inefficiencies in resource allocation also result in reduced system performance. Underutilized resources could have been effectively used to enhance system performance, while over-provisioning can cause performance bottlenecks or inefficient resource use, impacting the overall stability and scalability of the system.

Moreover, such resource wastage directly affects the **return on investment** (**ROI**) for Kubernetes deployments. Added costs due to underutilization or over-provisioning detract from the potential savings and operational efficiency that Kubernetes promises, diminishing the value derived from the investment in such systems.

## Security vulnerabilities and data breach possibilities

The presence of security vulnerabilities within a Kubernetes environment poses a significant risk, potentially exposing the system to data breaches and compromising sensitive information. The vast nature of Kubernetes, with its diverse interactions, heightens the vulnerability landscape, creating multiple entry points for potential security threats.

An exploited vulnerability can lead to unauthorized access, data leaks, or service disruptions, significantly compromising the confidentiality, integrity, and availability of critical data and services. Breaches in the Kubernetes environment can result in the exposure of sensitive information, leading to financial losses, legal repercussions, and damage to the organization's reputation.

The decentralized nature of the environment often leads to difficulties in enforcing consistent security controls across the entire infrastructure. Interconnections between microservices and containers also make it challenging to identify and address vulnerabilities in a timely manner.

The implications of security vulnerabilities expand beyond the system itself. Breaches in a Kubernetes environment not only impact the internal infrastructure but also potentially affect customers, partners, and stakeholders, eroding their trust and confidence in the organization.

Employing encryption techniques, adopting strict access controls, continuously monitoring for potential vulnerabilities, and ensuring regular security patches and updates are essential. A proactive approach to security, alongside ongoing security awareness training for personnel, is crucial in fortifying the system against potential threats.

## Hindrance to innovation and development

Hindrance to innovation and development within a Kubernetes environment stifles the evolution and progress of the system, creating impediments that significantly impact the organization's ability to adapt and innovate.

The intricacies of managing and optimizing a Kubernetes infrastructure can divert the focus and resources of development teams, limiting their capacity to innovate and create. As teams grapple with the complexities of the system, their time and efforts are often directed toward maintenance, troubleshooting, or understanding the Kubernetes architecture instead of dedicating these resources to fostering new innovations and enhancements.

This often results in longer lead times for deploying new features or applications. This delay in deployment can impact the organization's ability to be agile and responsive to market demands. Prolonged development cycles not only hinder the timely delivery of new services or features but also impede the organization's competitive edge in a dynamic business landscape.

Slower innovation cycles might result in missed opportunities as the organization struggles to adapt to changing market needs and emerging technologies, thereby potentially losing out on market share and growth potential.

## Team productivity and collaboration challenges

The intricacies of Kubernetes require deep expertise, which can create silos within teams due to specialized knowledge. This siloed approach can lead to difficulties in knowledge sharing, hindering cross-team collaboration and efficient problem-solving. The compartmentalization of knowledge and responsibilities can impede the collective efforts necessary for effective system management.

Kubernetes often requires a significant learning curve for team members, affecting their productivity and efficiency. This learning curve, coupled with the constantly evolving nature of Kubernetes, can lead to a drain on resources and time, affecting the overall productivity of teams. This diversion of time and resources toward understanding and managing Kubernetes complexities can take away from more strategic and productive tasks.

Furthermore, challenges in collaboration and communication across teams can impact the system's overall efficiency. Inconsistencies in communication channels or difficulties in sharing knowledge can slow down decision-making processes and troubleshooting efforts, leading to delays in problem resolution and system optimization.

Encouraging knowledge sharing, cross-team collaborations, and the implementation of comprehensive training programs can help alleviate knowledge silos and streamline team efforts.

## Business reputation and customer trust impacts

Security vulnerabilities, operational disruptions, or service degradation that can arise within a Kubernetes environment directly impact the business reputation and erode customer trust.

The consequences of business reputation and customer trust impacts can be far-reaching. Customers, partners, and stakeholders may lose trust in the organization's ability to safeguard their data and privacy, leading to a loss of confidence in the provided services. This loss of trust can translate into decreased customer retention rates and dissuade potential clients from engaging with the organization.

Moreover, operational disruptions or service degradation due to issues within the Kubernetes environment can adversely impact the customer experience. When services are unreliable or exhibit inconsistencies, customers can become frustrated, leading to a negative perception of the organization. Poor experiences can result in customer dissatisfaction, increased support requests, and, in some cases, customer churn.

The implications for the organization's reputation are profound. A damaged business reputation impacts brand loyalty, market positioning, and the overall credibility of the organization. In an increasingly competitive business landscape where trust and reputation are crucial differentiators, negative perceptions arising from issues within the Kubernetes environment can significantly impact the success and growth of the organization.

Prioritizing robust security measures, regular audits, and swift response to issues, along with proactive customer communication during disruptions, are essential. Establishing transparent and reliable customer communication channels can help mitigate negative perceptions and maintain customer trust.

# Summary

This chapter continued our exploration by delving deeper into practical aspects of identifying prevalent anti-patterns within Kubernetes ecosystems.

We meticulously examined 10 of the most common anti-patterns observed in the Kubernetes ecosystem. Each anti-pattern was dissected, accompanied by real-world consequences and explanations, enabling you to grasp the nuances of these deceptive patterns. These insights were aimed not only at theoretical understanding but also at aiding in recognizing these deceptive patterns within your own systems.

Moving further into the narrative, it portrayed real-world consequences that arise when these anti-patterns are allowed to persist within Kubernetes environments. It vividly illustrated tangible impacts, such as system failures, security vulnerabilities, operational disruptions, and financial losses, resulting from these anti-patterns. This section aimed to emphasize the critical importance of actively recognizing and mitigating these anti-patterns to ensure the stability and resilience of your Kubernetes setup.

Having traversed through the practical aspects of identifying common Kubernetes anti-patterns, we are now better equipped to navigate the complex terrain of Kubernetes anti-patterns. Armed with insights from their real-world implications, characteristics, and broader impact, our journey continues with a mission to actively recognize, address, and ultimately overcome these concealed challenges within Kubernetes environments.

In the next chapter, we'll explore the causes and consequences of Kubernetes anti-patterns, unraveling their root causes and tracing their influence while emphasizing the value of understanding for informed decision-making and proactive strategies.

# 3

# Causes and Consequences

This chapter offers an insightful exploration into the root causes and extensive consequences of Kubernetes anti-patterns, highlighting their impact on system operations. It provides a detailed analysis of Kubernetes' historical development, addressing common misconceptions and knowledge gaps among practitioners. The text emphasizes the significant role of architectural and organizational factors in Kubernetes deployments and the importance of human elements such as skills, training, and communication in effective management. Additionally, it assesses the influence of tooling and technology choices on the operational efficiency of Kubernetes environments. Overall, this chapter aims to provide a comprehensive understanding of Kubernetes anti-patterns, focusing on proactive strategies to ensure operational stability and functionality.

We will cover the following topics in this chapter:

- Unpacking the root causes of Kubernetes anti-patterns
- Tracing Kubernetes anti-pattern influence
- The value of understanding anti-pattern causes

## Unpacking the root causes of Kubernetes anti-patterns

Understanding the root causes of anti-patterns in Kubernetes is key to mastering the platform's complexities and optimizing its use. This exploration sheds light on the intricate factors that contribute to operational challenges, from the evolution of Kubernetes and its impact on current practices to the nuances of organizational dynamics, technical skills, and tool choices.

### Defining root causes within Kubernetes

One critical aspect demands our attention: the root causes of anti-patterns. These are not just surface-level glitches, but rather the foundational issues that trigger a chain of operational challenges. To address these effectively, we must understand what exactly constitutes a root cause in the Kubernetes ecosystem.

Root causes in Kubernetes are often buried under layers of complexity. For instance, the overutilization of resources might initially appear as a straightforward issue of workload mismanagement. But the actual root cause might be traced back to a fundamental misunderstanding of Kubernetes' resource management features, such as requests and limits for pods and containers.

In Kubernetes environments, effectively distinguishing between symptoms and root causes is vital. It's the difference between applying a temporary fix and solving the problem at its core. This distinction not only resolves the immediate issue but also enhances the long-term health and stability of the system.

It is a blend of investigative work and technical acumen. It involves dissecting the system architecture and operational practices and then applying a systematic approach to problem-solving. Tools such as log analyzers, monitoring systems, and Kubernetes-specific diagnostics are invaluable in this process.

The importance of root cause analysis is highlighted in real-world scenarios. Consider a situation where a Kubernetes deployment suffers from frequent downtime. Simply restarting services or reallocating resources might offer a temporary reprieve. However, a deeper investigation might reveal a more complex issue, such as flawed deployment strategies or network policy misconfigurations. Addressing these root causes directly leads to a more sustainable and effective solution.

The following table shows how specific anti-patterns can be traced to deeper underlying causes:

| Kubernetes Anti-Pattern | Potential Root Cause |
|---|---|
| Resource Overutilization | Misconfigured resource limits and requests |
| Frequent Downtime | Flawed deployment strategies |
| Security Vulnerabilities | Inadequate security policies and practices |
| Scalability Issues | Architectural limitations |
| Inefficient Workload Distribution | Poor understanding of Kubernetes scheduling |

Table 3.1 – Kubernetes anti-patterns and their root causes

When undertaking root cause analysis in Kubernetes, it's essential to adopt a systematic and comprehensive approach. This involves several key steps:

1. **Incident logging and preliminary analysis**: Begin by documenting the incident thoroughly. Gather all relevant data, including logs, metrics, and user reports. This stage involves identifying the symptoms of the issue.

2. **Data collection and tool utilization**: Utilize Kubernetes-specific tools for deeper insights. This might include log analyzers such as ELK Stack (Elasticsearch, Logstash, Kibana), monitoring tools such as Prometheus, and Kubernetes diagnostic tools such as `kubectl` for inspecting running pods and workloads.

3. **Hypothesis formulation**: Based on the preliminary data, formulate hypotheses about potential root causes. This stage is speculative but guided by the data collected and the practitioner's knowledge of Kubernetes operations.

4. **Testing hypotheses**: This involves experimenting within the Kubernetes environment to validate or refute each hypothesis. It might include replicating scenarios, adjusting configurations, or simulating workloads.

5. **Involving cross-functional teams**: Given the complexity of Kubernetes, involve cross-functional teams in the RCA process. This might include developers, system architects, and operations teams, each bringing a unique perspective to the problem.

6. **Identifying the root cause**: After thorough testing and collaboration, narrowing down to the specific root cause requires a blend of analytical techniques:

    - **Correlation analysis**: Linking data patterns from logs, metrics, and alerts to identify potential causes

    - **Comparative analysis**: Contrasting configurations and metrics with those of correctly functioning systems

    - **Causal inference**: Establishing cause and effect relationships between observed changes and issues

    - **Expert consultation**: Drawing on the knowledge and experiences of seasoned Kubernetes professionals

    - **Backtracking**: Tracing the sequence of events or changes leading up to the problem using tools such as Kubernetes audit logs

7. **Documenting and sharing findings**: Once the root cause has been identified, document the findings comprehensively. Share this documentation with relevant teams to ensure awareness and prevent recurrence.

8. **Implementing corrective actions**: Finally, implement the necessary corrective actions. This could involve changes in configuration, updates in deployment practices, or revisions in the architectural approach.

9. **Review and continuous improvement**: Post-RCA, review the process for any learning points and potential areas of improvement. This helps refine the RCA process for future incidents.

A successful RCA in Kubernetes is not just a technical exercise; it's a strategic approach that combines technical expertise, collaborative problem-solving, and a commitment to continuous learning. By mastering this approach, Kubernetes practitioners can transform operational challenges into opportunities for optimization and growth, leading to more stable and efficient Kubernetes environments.

# Historical perspectives on Kubernetes development

The historical evolution of Kubernetes offers a crucial backdrop for understanding the root causes of prevalent anti-patterns in its use today. Far from being a mere retrospective, this historical lens is vital for comprehending how Kubernetes has developed over time and why certain anti-patterns have become entrenched.

Kubernetes, born out of Google's internal Borg system and later donated to the **Cloud Native Computing Foundation** (**CNCF**), was developed to orchestrate containerized applications at scale. In its early days, the focus was on creating a robust platform capable of managing complex applications efficiently. This was a time when Kubernetes was primarily centered around scalability and the automation of deployment, scaling, and operations across clusters of hosts. The platform excelled at managing stateless applications, which constituted the majority of workloads during its early stages.

However, as Kubernetes began to gain popularity, its feature set expanded. This expansion, while enriching the platform, also brought with it a complexity that had not been encountered previously. Kubernetes started to support stateful applications and a wider array of workload types. Each new feature, from persistent storage to network policies, introduced new dimensions of configuration and management. This growth, while a testament to Kubernetes' versatility, also began to sow the seeds of what would become common anti-patterns.

The development of Kubernetes was significantly shaped by its active community. The contributions from various organizations and individuals brought diverse perspectives and use cases to the table. This community-driven development was a double-edged sword: while it propelled rapid innovation and adaptation, it also introduced a level of inconsistency and complexity to Kubernetes' evolution. Practices and patterns that were effective in specific contexts were sometimes adopted more broadly without sufficient vetting for their general applicability, leading to missteps and inefficiencies.

With the proliferation of Kubernetes deployments came the rise in instances of suboptimal usage – the so-called anti-patterns. These often stemmed from the platform's inherent complexity and a lack of established best practices in the early stages of its adoption. Many users, particularly those without the resources and expertise on the scale of companies such as Google, found themselves inadvertently adopting practices that were ill-suited to their specific needs or contexts.

Understanding this historical context is critical in recognizing why certain anti-patterns exist in the world of Kubernetes. It illuminates the impact of early design decisions, the challenges brought on by rapid feature evolution, and the influence of a diverse community on the Kubernetes landscape. This understanding is not just about identifying the root causes of existing anti-patterns; it's about gaining insights that can help with anticipating and navigating potential future challenges as Kubernetes continues to evolve and shape the landscape of container orchestration.

# Common misconceptions and knowledge gaps

In the quest to understand and rectify Kubernetes anti-patterns, we must confront a critical element that often acts as a catalyst for these issues: the misconceptions and knowledge gaps prevalent among practitioners. This part of our journey delves into these misunderstandings, unraveling how they contribute to the root causes of common Kubernetes pitfalls.

## Misconception 1 – Kubernetes as a universal solution

One of the most pervasive misconceptions is viewing Kubernetes as a one-size-fits-all solution. Originating from its widespread popularity and success stories, this view often leads to its adoption in scenarios where it may not be the most suitable choice, inadvertently setting the stage for anti-patterns. Understanding that Kubernetes, while powerful, is not always the optimal solution for every use case is crucial in avoiding misapplication.

## Misconception 2 – overestimation of automation

Kubernetes is often heralded for its automation capabilities, but there's a tendency to overestimate what it can automate out of the box. This overestimation can lead to underinvestment in necessary customizations and manual oversight, resulting in misconfigured environments and operational issues. Recognizing the balance between automation and manual intervention is key to harnessing Kubernetes effectively.

## Misconception 3 – simplistic views on Kubernetes' complexity

Many practitioners tend to underestimate the complexity of Kubernetes, which can lead to significant challenges. New adopters, often influenced by its user-friendly frontend and high-level abstractions, may overlook the intricacies involved in its setup and maintenance. This gap in understanding can lead to simplistic implementations, which fail to account for the nuances of a robust Kubernetes deployment.

## Misconception 4 – equating Kubernetes with its tools

The line between Kubernetes and the various tools and add-ons used in conjunction with it is often blurred. Tools such as Helm, Istio, or Prometheus, while valuable, are distinct from Kubernetes itself. This blurring can result in an overdependence on these third-party tools, overshadowing the need for a fundamental understanding of Kubernetes mechanics, and may lead to configurations that fail to capitalize on the core strengths of Kubernetes.

## Misconception 5 – the fallacy of "set and forget"

There's often a belief that once Kubernetes is set up, it requires minimal maintenance. This *set and forget* mindset overlooks the dynamic nature of Kubernetes environments and the ongoing monitoring, updating, and optimization they require. Such an attitude can lead to outdated systems, security vulnerabilities, and performance degradation.

### Knowledge gap 1 – inadequate understanding of Kubernetes architecture

A significant knowledge gap often lies in a comprehensive understanding of Kubernetes architecture. The nuances of its components – pods, services, deployments, and more – and their interactions are sometimes not fully grasped, leading to suboptimal configurations that evolve into anti-patterns.

### Knowledge gap 2 – misunderstanding Kubernetes networking

Networking in Kubernetes is a complex area that is frequently misunderstood. Concepts such as network policies, service meshes, and ingress and egress rules can be daunting. An incomplete grasp of these concepts often results in networking-related anti-patterns, impacting the performance and security of applications.

### Knowledge gap 3 – overlooking security best practices

Security in Kubernetes is paramount but often inadequately addressed due to knowledge gaps. The nuances of **role-based access control** (**RBAC**), secret management, and network security are areas where a lack of understanding can lead to serious vulnerabilities.

### Knowledge gap 4 – containerization versus Kubernetes optimization

A critical knowledge gap is the distinction between containerization and Kubernetes optimization. Simply containerizing applications doesn't automatically translate to optimized Kubernetes performance. An in-depth understanding of how Kubernetes orchestrates these containers, manages resources, and ensures high availability is vital for optimal deployment.

### Knowledge gap 5 – underappreciating the importance of observability

Observability (monitoring, logging, and tracing) in Kubernetes is often underappreciated. This oversight can lead to scenarios where issues go undetected or are identified too late. A comprehensive understanding of observability practices is crucial for proactive system health management.

To prevent and rectify Kubernetes anti-patterns, education and continuous learning are essential. Practitioners should seek to understand the platform's limitations and strengths, invest time in comprehending its complexities, and stay updated on best practices, particularly in areas such as networking and security.

## Architectural and design pitfalls

Grasping the nuances of Kubernetes architecture and design is a vital component for any practitioner. These aspects are not just mere technicalities; they represent a labyrinth of decisions and strategies that, if not navigated carefully, can lead to significant challenges in the Kubernetes environment. The nature of these challenges is often deep-rooted, stemming from foundational choices made during the initial phases of Kubernetes setup and its ongoing development.

The journey into Kubernetes architecture often begins with an enthusiasm for its capabilities. However, one of the first pitfalls many encounter is over-engineering. It's a common scenario: in an attempt to leverage the power of Kubernetes, there's a tendency to create overly complex systems. These systems, with their multiple layers and intricate components, can become unwieldy, difficult to manage, and prone to errors. The principle of simplicity is key here. An unnecessarily complex architecture not only hampers operations but also obscures the root causes when problems arise. The challenge lies in finding the balance between harnessing Kubernetes' robust features and maintaining a system that is manageable and not overburdened by its complexity.

Another critical area is cluster sizing and scalability. This aspect of Kubernetes architecture is akin to walking a tightrope. On one side is the risk of underestimating the necessary size of the cluster, leading to resource shortages, performance bottlenecks, and a system that groans under the weight of its workloads. On the other side is the danger of overestimating, resulting in resource wastage and unnecessary expenditures. Additionally, scalability planning is often overlooked or underestimated. Kubernetes environments must be designed with future growth in mind; failing to do so can result in a system that's unable to cope with increasing demands, thus negating one of Kubernetes' principal benefits.

The design of pods and services requires careful contemplation. Kubernetes shines in its orchestration capabilities, but missteps in pod and service design can quickly diminish its advantages. For instance, cramming too many containers into a single pod, or poorly defining the boundaries of services, can lead to decreased performance and escalated complexity. Each container, pod, and service needs to be thoughtfully configured to ensure they collectively work toward enhancing performance, not detracting from it.

Addressing stateful components within Kubernetes, primarily known for managing stateless applications, introduces its own set of challenges. Incorporating elements such as databases demands a strategic approach to stateful sets and persistent volumes. Mismanagement here can lead to data persistence issues that impact the reliability and effectiveness of the entire system. Ensuring data integrity and availability in a predominantly stateless environment requires a deep understanding of Kubernetes' storage capabilities and a meticulous approach to their implementation.

A pitfall that is often not given enough consideration until it's too late is disaster recovery and high availability. Designing a Kubernetes system without integrated failover mechanisms and a solid backup strategy leaves it exposed to potential failures and downtimes. Especially in production environments, the absence of robust disaster recovery planning can have catastrophic consequences. High availability and disaster recovery need to be woven into the fabric of the Kubernetes architecture from the beginning.

Finally, the integration of security within the Kubernetes architecture is an area that is frequently mishandled. Often treated as an afterthought, security needs to be a foundational component of the Kubernetes design. This includes everything from network segmentation to the effective use of Kubernetes' RBAC and securing intra-cluster communications. An architecture that overlooks these elements can be vulnerable to a range of security threats.

## Organizational dynamics and their effects

The way an organization is structured, how it makes decisions, its cultural orientation, the level of skills and training, and its approach to resource allocation collectively shape the Kubernetes landscape within that organization.

At the heart of many Kubernetes deployments is the structure of the organization. Traditional, siloed structures, where departments operate as separate entities, often lead to fragmented Kubernetes practices. When teams work in isolation, without a platform for cross-collaboration or shared learning, disparities in deployment practices and configurations emerge. This disjointed approach can inadvertently breed anti-patterns as teams might adopt different strategies or levels of Kubernetes maturity. The key is to foster a unified approach, where consistent practices and knowledge sharing across teams are emphasized.

Decision-making processes within an organization play a critical role in the successful adoption and management of Kubernetes. In environments where decisions regarding Kubernetes are made at higher levels without consulting the actual users or administrators, there's a risk of misalignment with the operational realities. This top-down decision-making can lead to the adoption of tools or practices that do not align with the technical needs or capabilities of the team, thereby laying the groundwork for anti-patterns to flourish.

The cultural landscape of an organization is another cornerstone in Kubernetes adoption. An organization that resists change or new technologies might find itself struggling to embrace the agile, iterative nature required for Kubernetes. In contrast, a culture that encourages innovation, experimentation, and continuous learning can be a fertile ground for successful Kubernetes strategies. Such a culture empowers teams to explore, learn, and adopt practices that are most effective, reducing the likelihood of falling into anti-patterns.

The skill levels within the organization and the emphasis placed on training are equally crucial. Kubernetes is a complex system with a steep learning curve, and a team that isn't well-versed in its intricacies is more likely to make mistakes in configuration and deployment. Organizations that invest in ongoing training and skill development create a foundation for more robust and efficient Kubernetes usage, preventing common pitfalls that lead to anti-patterns.

Resource allocation and prioritization within an organization can significantly impact Kubernetes management. When Kubernetes initiatives are under-resourced or not given sufficient priority, the result can be hastily executed deployments, inadequate testing, and poor configurations – all of which are breeding grounds for anti-patterns. On the other hand, adequate resource allocation and prioritization enable thorough planning, robust testing, and effective management, supporting a healthy Kubernetes environment.

Organizational changes, such as restructuring, mergers, or shifts in strategic direction, can also have profound effects on Kubernetes environments. Such changes can disrupt established practices and bring new challenges. Navigating these changes with a focus on maintaining Kubernetes best practices is essential to prevent destabilizing the existing environment and introducing anti-patterns.

# The human element – skills, training, and communication

The human element plays an indispensable role. This aspect, often overshadowed by the technicalities, is a cornerstone in either paving the way to success or leading to the emergence of anti-patterns in Kubernetes deployments.

In the Kubernetes ecosystem, the skillset of the team is of paramount importance. It's a domain where understanding extends beyond the mere basics of containerization; it demands a comprehensive grasp of Kubernetes' multifarious functionalities, such as networking, storage, and security. Without this depth of knowledge, teams are prone to fundamental errors in deployment and management, leading to inefficiencies and vulnerabilities. For instance, a lack of understanding about Kubernetes networking can lead to poorly configured services, while insufficient knowledge about security practices might leave the system exposed to threats. Therefore, ensuring that your team possesses a broad and deep skill set is essential to avoid these pitfalls.

However, skills alone are not enough. The field of Kubernetes is dynamic and constantly evolving, with new features and changing best practices emerging. This is where continuous training plays a critical role. Organizations must not only provide initial training but also invest in ongoing education to keep their teams up to speed with the latest advancements in Kubernetes. Such training should encompass more than just the technicalities; it should also include operational, security, and scalability best practices. A team that is regularly trained is more capable of not only navigating the complexities of Kubernetes but also leveraging its full potential.

The significance of communication within and across teams in managing Kubernetes cannot be overstated. In environments where communication is fragmented, it's common to see disjointed strategies in Kubernetes implementation. This can result in a patchwork of practices across different teams or departments, often leading to inconsistent and suboptimal deployments. Effective communication ensures a unified approach, aligning strategies and fostering a culture where knowledge and insights are shared freely. This ensures that everyone involved with Kubernetes is aligned with the organizational goals and is working in concert, which is vital in preventing misalignments and inefficiencies.

By focusing on building a team with the right skills, ensuring continuous and comprehensive training, and fostering open and effective communication, organizations can create a robust foundation for Kubernetes deployment and management. This approach not only minimizes the risk of anti-patterns but also enables teams to fully harness the capabilities of Kubernetes, turning challenges into opportunities for innovation and growth.

## Tooling and technology choices

Selecting and integrating the right tools and technologies is a critical process in managing Kubernetes environments. These decisions are not just routine selections; they shape the very fabric of how Kubernetes operates, affecting operational efficiency, scalability, and security. This process involves navigating a landscape filled with diverse tools, each promising specific enhancements and optimizations to container orchestration.

Faced with such a wide array of choices, the key challenge is to discern which tools align best with the deployment's specific needs and goals. This task requires performing an in-depth analysis of how each tool fits into the existing system, understanding the resources and time needed to learn and implement them effectively, evaluating the support available within the Kubernetes community, and considering their long-term maintainability.

The temptation to adopt new, advanced tools is common, but it must be tempered with a comprehensive assessment of their appropriateness for the given Kubernetes environment. For example, incorporating a complex service mesh solution might seem like a progressive move, but if it's not necessary for the specific use case, it can add unnecessary complexity to operations. Similarly, selecting a monitoring tool that doesn't align well with Kubernetes can lead to significant blind spots, undermining the environment's management and monitoring capabilities.

Beyond auxiliary tools, the broader technology stack that integrates with Kubernetes also demands careful consideration. This includes storage, networking, and security solutions that need to work harmoniously with Kubernetes' architecture. Incompatible choices in this technology stack can lead to issues such as data persistence challenges or network traffic bottlenecks, which can severely impact the performance and reliability of the entire system.

Moreover, the way these tools and technologies are implemented within Kubernetes is critical. Even the most robust tools will fall short if they are not configured correctly and optimized for the Kubernetes environment. Misconfigurations or inefficient setups can negate the benefits of these tools, leading to inefficiencies and vulnerabilities. Therefore, it's essential to have a deep understanding of both the tools and Kubernetes to ensure that the integration of these tools enhances the Kubernetes ecosystem.

Thus, the journey of selecting and implementing tools and technologies in Kubernetes requires careful thought and informed decisions. It's a process that balances the excitement of new technological advancements with the practical needs and specifics of the Kubernetes environment. Through thoughtful selection and meticulous implementation, practitioners can create a Kubernetes ecosystem that is not just functional but also optimized for efficiency, security, and scalability.

## Tracing Kubernetes anti-pattern influence

Tracing the influence of Kubernetes anti-patterns is crucial in uncovering how they subtly, yet significantly, impact the platform's operational efficiency and effectiveness. This section delves into the various ways in which these anti-patterns, often overlooked or misunderstood, can distort and challenge the norms of Kubernetes use. From misaligned practices to misconfigured settings, understanding the breadth and depth of these anti-patterns provides valuable insights into the complexities of Kubernetes and how best to navigate them for optimal performance.

# Subtle shifts in development culture

Kubernetes anti-patterns, often emerging from misapplications or misunderstandings of Kubernetes' capabilities, subtly influence the development culture in various ways. One of the most significant shifts is the emergence of an over-reliance on Kubernetes' automated features. Developers might begin to depend heavily on Kubernetes for handling various aspects of application deployment and scaling, assuming that the platform will automatically resolve any configuration or architectural inefficiencies. This misplaced confidence can lead to a neglect of core software engineering principles as teams increasingly lean on Kubernetes to *fix* suboptimal practices.

Another cultural shift induced by Kubernetes anti-patterns is the centralization of expertise and knowledge. As teams encounter more complex Kubernetes environments, exacerbated by anti-patterns such as inappropriate use of resources or misconfigured services, a small group within the team often becomes the de facto Kubernetes experts. This situation creates knowledge silos, where few team members hold most of the understanding related to Kubernetes operations. Consequently, other team members may feel disconnected from the Kubernetes-related aspects of projects, leading to a decline in overall team efficiency and a potential increase in errors due to miscommunication or lack of understanding.

Kubernetes anti-patterns also tend to encourage a culture of reactive rather than proactive problem-solving. When teams repeatedly encounter issues stemming from these anti-patterns, such as resource contention or service outages due to improper load balancing, the focus shifts to putting out fires rather than preventing them. This reactive approach can become ingrained in the team's culture, prioritizing immediate solutions over a thorough analysis and understanding of the underlying problems. This mindset often leads to a cycle of quick fixes, which, while providing temporary relief, do not address the root causes of issues.

Furthermore, the presence of Kubernetes anti-patterns can lead to a culture of complacency regarding continuous improvement and learning. With the complexity introduced by these anti-patterns, team members might feel overwhelmed or resigned to the notion that the Kubernetes environment is inherently problematic. This attitude can stifle innovation and discourage team members from seeking out new and better ways to utilize Kubernetes effectively. It can also lead to a stagnation of skills as team members become less inclined to update their knowledge or explore the evolving best practices in Kubernetes usage.

Additionally, these anti-patterns can subtly shift the team's approach to risk management and testing. In a healthy Kubernetes environment, teams would typically conduct thorough testing, including load testing, failover scenarios, and recovery procedures. However, when anti-patterns are prevalent, there's often a false sense of security that Kubernetes will manage these aspects effectively. As a result, teams might skip comprehensive testing, leading to vulnerabilities in the system that are only discovered when they fail in production environments.

The adoption of Kubernetes anti-patterns can subtly influence the team's approach to architecture and design. The allure of Kubernetes' features might lead teams to design systems that are overly complex and intertwined with Kubernetes-specific functionalities. This over-reliance on Kubernetes can make the system rigid and less adaptable to changes, locking the architecture into patterns that are difficult to scale or modify.

When faced with the complexities and challenges of an improperly managed Kubernetes environment, team members might become less likely to collaborate effectively. This increased complexity can lead to a lack of transparency and shared understanding of the system, making it challenging for team members to work together efficiently on solving problems or developing new features.

## Workflow disruptions and inefficiencies

The daily workflows of development teams are deeply influenced by Kubernetes anti-patterns, which, although technical in their origin, lead to a range of inefficiencies and disruptions. The impact of these patterns is not straightforward; rather, it manifests as a web of interlinked challenges, each affecting the team in different, often unexpected ways. Navigating this complexity becomes a crucial part of managing a Kubernetes environment effectively.

When Kubernetes is not leveraged correctly, one of the most immediate impacts is on the deployment and management of applications. Anti-patterns, such as misconfigured resource limits or inappropriate use of Kubernetes objects, can result in frequent deployment failures. Teams are pulled away from their planned tasks to address these urgent issues, leading to a cycle of reactive problem-solving that disrupts regular workflows and diminishes productivity.

Troubleshooting and maintenance become increasingly complex due to these anti-patterns. For instance, overly complex networking configurations or the excessive use of custom resource definitions can obscure the underlying causes of issues. This complexity forces teams to spend significant time untangling these problems, which delays other critical work and advances in project development.

Inefficient resource utilization is another consequence of Kubernetes anti-patterns. Practices such as neglecting to set appropriate resource limits can lead to excessive or insufficient resource allocation. This not only affects application performance but also leads to higher operational costs and wasted resources, something that requires frequent adjustments and monitoring.

Scaling applications effectively becomes a challenge with the presence of Kubernetes anti-patterns. Misconceptions around scaling strategies can result in applications that fail to scale properly under varying loads. As a result, teams often find themselves manually managing application scaling, which is time-consuming and disrupts their focus on other developmental aspects.

Collaboration and communication within teams are also impacted. Misconfigured Kubernetes environments can lead to misunderstandings and increased communication overhead as team members struggle to clarify configurations and deployment strategies. This inefficient communication slows down the development process and can lead to frustration and decreased morale.

The integration of **continuous integration and continuous deployment (CI/CD)** processes can be hindered by Kubernetes anti-patterns. Suboptimal configurations or complex deployment strategies can cause CI/CD pipelines to fail frequently, delaying software delivery and diverting attention from feature development to pipeline troubleshooting.

System reliability and predictability suffer in the presence of Kubernetes anti-patterns. Improperly managed and configured systems are more prone to failures and unpredictable behavior, necessitating a constant state of vigilance from teams. This unpredictability hampers the ability to plan and execute work effectively, leading to a more chaotic and stress-filled work environment.

## Altered deployment and operational metrics

Common pitfalls in Kubernetes can subtly and significantly impact the metrics that are used throughout the Kubernetes ecosystem to evaluate deployment and operational efficiency. These often subtle shifts can alter the conventional perception of the ecosystem's performance.

Deployment frequency, a metric often equated with agility, can be misleading when anti-patterns are at play. An increase in deployment frequency might appear as a positive indicator, but it could also signify rushed releases, inadequate testing, or a lack of readiness for production, leading to instability and potential downtime in the ecosystem.

Change failure rate experiences a similar shift. While Kubernetes can mask immediate deployment failures, resulting in a seemingly improved change failure rate, this can hide deeper issues. Problems such as configuration drift or resource contention, which are not immediately evident, can gradually erode the system's stability and resilience, impacting the long-term health of the Kubernetes ecosystem.

The **mean time to recovery (MTTR)** is another metric that's affected by Kubernetes anti-patterns. The platform's ability to quickly roll back changes and restore previous states might create an impression of a resilient system. However, this can also prevent teams from addressing underlying causes of failures, leading to a cycle of recurring issues that can destabilize the ecosystem over time.

Furthermore, the necessity to monitor new, Kubernetes-specific operational metrics becomes evident. Metrics related to container orchestration, pod performance, and node health become crucial. Properly tracking and interpreting these metrics requires a deep, nuanced understanding of Kubernetes, and failing to do so can lead to misjudged system performance and health.

The complexity that's introduced by these anti-patterns also complicates the interpretation of operational metrics. Teams must navigate a more intricate data landscape, understanding not just the metrics themselves but also how Kubernetes' features and anti-patterns might be influencing them. This complexity makes it challenging to draw accurate conclusions about the system's performance and to make informed decisions regarding improvements and resource management.

Therefore, in the Kubernetes ecosystem, the influence of anti-patterns extends to altering key operational metrics, necessitating a more refined approach to measurement and analysis to ensure a true understanding of the system's health and efficiency.

## Increased monitoring noise and alert fatigue

Kubernetes anti-patterns can inadvertently cause a surge in monitoring noise and alert fatigue, which poses a significant challenge for teams overseeing complex, dynamic systems. This scenario unfolds as the system begins generating an overwhelming number of alerts and logs, many of which might be trivial or misleading, yet they still require assessment and management.

Anti-patterns such as misconfigured resource thresholds or health checks, and the inappropriate use of alerts, often lead to a flood of notifications. For example, if the alert thresholds are set too sensitively or without proper context for the specific Kubernetes environment, teams may find themselves inundated with alerts for normal system behavior or minor deviations. This constant influx of alerts can create a background noise, making it difficult to discern genuinely critical issues.

The consequence of this incessant stream of alerts is a growing desensitization among team members. Continuously bombarded with notifications, individuals start to experience alert fatigue, where they might overlook or undervalue significant alerts, mistaking them for routine false positives. This condition is dangerous as it allows real, potentially system-critical issues to go unnoticed or unaddressed, increasing the risk of major system failures or performance issues.

This issue extends to the realm of log management as well. Kubernetes environments, especially those with embedded anti-patterns, tend to produce an extensive amount of log data. When this data is inflated with entries stemming from these anti-patterns, it not only strains the storage and processing capacities but also complicates the task of sifting through logs to extract actionable insights. Teams are forced to spend disproportionate amounts of time filtering through this data, trying to identify relevant information amid a sea of logs.

Excessive alerts and logs demand a more thoughtful approach to monitoring. Teams are compelled to refine their alerting systems, ensuring that the thresholds and conditions for alerts truly reflect critical issues in the system. Similarly, advanced log management strategies become necessary, typically those that utilize tools and techniques that can effectively parse through large volumes of data, filter out the noise, and highlight key information that requires attention.

It is crucial for teams to continuously evaluate and adjust their monitoring setups to ensure they are both effective in catching real issues and efficient in not overwhelming the team with unnecessary noise. This careful management of monitoring systems is vital for maintaining the health and stability of Kubernetes environments, ensuring that teams can focus on genuine issues and maintain optimal system performance.

## Degradation of service reliability

The degradation of service reliability in a Kubernetes environment is often a direct result of various anti-patterns in configuration and usage. When resources are misallocated or improperly managed, it can lead to services being either resource-starved or over-provisioned. The former results in frequent crashes or slowdowns under load, while the latter leads to wasted resources and increased operational costs. This mismanagement directly affects the dependability of services, with them failing to meet user expectations and service-level agreements.

Configuring liveness and readiness probes inaccurately can also significantly impact service stability. If these probes are not tuned correctly, Kubernetes may terminate healthy containers unnecessarily or fail to restart those that are malfunctioning. This can result in increased downtime or poor service response as Kubernetes struggles to accurately assess the state of running containers.

Network configuration within Kubernetes is crucial for service reliability, especially in microservices architectures where communication between services is key. Issues, such as misconfigured network policies or service ingress, can lead to services being unreachable or experiencing erratic network behavior. This can manifest as increased latency, packet loss, or total service outages, further eroding the reliability of the system.

Load balancing and auto-scaling rules that are not optimally configured can disrupt service reliability as well. Imbalanced traffic distribution can overload some parts of the system while underutilizing others. Auto-scaling that responds too slowly to changes in demand, or scales down too aggressively, can leave services unable to cope with user requests effectively, impacting both availability and user experience.

Managing stateful applications in Kubernetes introduces additional complexities that can affect service reliability. Missteps in handling data persistence, StatefulSets, or persistent volumes can lead to data loss, corruption, or inconsistency, particularly during pod restarts or scaling operations. These issues directly challenge the integrity and reliability of the services relying on this data.

Visibility into the system through monitoring and logging is crucial for maintaining service reliability. A lack of adequate monitoring can obscure the root causes of reliability issues, making them difficult to diagnose and resolve. Without clear insights into how services are performing and how resources are being utilized, teams may struggle to identify and address the configurations contributing to service instability.

In essence, multiple factors stemming from anti-patterns in Kubernetes can contribute to a decline in service reliability. From resource mismanagement and probe configuration errors to networking challenges and stateful application complexities, these issues need to be paid careful attention to ensure that services remain stable, responsive, and reliable.

## Complications in automation and orchestration

The complexities of automating and orchestrating tasks can lead to significant operational challenges. These challenges often stem from the delicate balance that must be maintained between automated processes and the need for tailored orchestration strategies. Let's take a closer look:

- **Challenges with over-reliance on automation**: Automation intends to ease the management of application deployment, scaling, and maintenance. However, teams may fall into the trap of *automation complacency*, a state where too much dependence is placed on automated processes. This over-reliance can be detrimental, particularly in scenarios that automated workflows are not equipped to handle. It often results in overlooked issues and inadequate resource management as manual intervention and customization are underutilized.

- **Tailoring orchestration to application needs**: Efficient orchestration in container management requires each application's unique requirements to be considered carefully. A common misstep occurs when the orchestration process is uniformly applied across diverse applications, disregarding their operational characteristics. Such a one-size-fits-all approach can lead to inefficient use of resources and diminished application performance. For example, imbalances in load balancing or inadequate pod deployment strategies can cause an uneven distribution of resources, directly impacting application effectiveness.

- **Complexities in network orchestration**: Orchestrating network configurations is a critical yet intricate aspect of container management. Network policies, which are essential for operational efficiency and security, need to be meticulously crafted. Improperly designed network policies can create excessive complexity, leading to difficult-to-manage interdependencies and unintentional access limitations. These issues not only degrade performance but also introduce substantial security risks.

- **Handling stateful applications**: Managing stateful applications presents unique challenges, especially since the platform is primarily optimized for stateless applications. Effectively managing StatefulSets and persistent volumes is crucial. Common errors include applying stateless strategies to stateful applications, which can result in significant data consistency issues and potential data loss, both of which pose major risks in environments where data integrity is critical.

- **Pitfalls in CI/CD processes**: Implementing CI/CD processes often emphasizes speed, sometimes at the expense of stability and thorough testing. This can lead to updates being deployed prematurely and unstable or untested code being introduced to production. Such practices compromise the reliability and efficiency of the system, leading to potential instability and disruptions.

Addressing these complexities requires a nuanced approach where automation benefits are balanced with necessary human insights, orchestration strategies are customized per application, and rapid deployment processes are aligned with the system's stability and security needs. Optimizing these aspects is crucial for preventing operational inefficiencies and the development of anti-patterns in container orchestration environments.

## Obstacles in performance tuning

Navigating the complexities of performance tuning in Kubernetes environments is often compounded by the presence of anti-patterns, which create a series of obstacles that hinder optimal performance. These anti-patterns, which are deeply rooted in various aspects of Kubernetes, can significantly distort the effectiveness of what is otherwise a robust and efficient system.

In the sphere of resource allocation, anti-patterns often emerge as either over or under-provisioning of resources. Kubernetes environments plagued by these anti-patterns struggle with efficiently distributing CPU and memory, leading to scenarios where some containers consume more resources than necessary, while others starve, resulting in erratic application performance. This mismanagement is particularly problematic in dynamic environments where workloads vary as the systems fail to adapt resource allocation in response to changing demands.

Load distribution in Kubernetes can be severely impacted by anti-patterns. Instead of an even spread of workloads across the cluster, these anti-patterns often cause an imbalanced distribution, overburdening certain nodes while leaving others underutilized. This not only strains the overburdened nodes, which can potentially lead to failures, but also signifies a gross inefficiency in utilizing the available infrastructure.

Network performance in Kubernetes is another area where anti-patterns can have a detrimental effect. Misconfigured network policies or inefficient networking strategies can lead to increased latency and reduced throughput. This is often a result of a lack of understanding of Kubernetes' networking capabilities or oversights in network design, leading to bottlenecks that impede the smooth operation of applications.

When it comes to storage, anti-patterns in Kubernetes manifest as poorly designed storage solutions. This can result in slow data access and create bottlenecks, particularly for stateful applications where fast and reliable storage access is critical. Such issues often arise from a mismatch between the chosen storage solution and the application's specific storage requirements.

Scalability, a key feature of Kubernetes, can be severely hampered by anti-patterns. Systems afflicted with these anti-patterns exhibit poor scalability, struggling to efficiently scale up or down in response to workload changes. This often stems from a lack of proper scalability planning or misconfigured auto-scaling parameters, leading to performance degradation during peak loads.

Monitoring and performance management, both of which are critical for maintaining system health, are often undermined by anti-patterns. Ineffective monitoring strategies or a lack of comprehensive monitoring can leave performance bottlenecks undetected and unaddressed. This leads to a reactive rather than proactive approach to performance management, where issues are only tackled once they have become critical.

Finally, caching, if not managed correctly, can become an anti-pattern in itself. An improper caching strategy can lead to inefficient memory usage, where either too much or too little cache is allocated, negatively impacting the system's overall performance. This is often a result of not understanding the application's specific caching needs or failing to adjust cache settings in line with those needs.

Each of these obstacles that are created by anti-patterns in Kubernetes performance tuning highlights the need for a thorough understanding of the system's capabilities and the pitfalls to avoid. This underlines the importance of a strategic approach to resource allocation, load balancing, network setup, storage management, scalability, monitoring, and caching to ensure the Kubernetes environment operates optimally.

# The value of understanding anti-pattern causes

Understanding what causes anti-patterns generates tangible, long-term benefits by empowering organizations to make informed decisions, anticipate issues, and build resilient, optimized Kubernetes environments. Let's delve into these valuable insights.

## Enabling predictive and preventive strategies

Organizations that grasp the nuances of anti-patterns in Kubernetes can set up systems that alert them to emerging issues. For instance, patterns leading to system strain in previous deployments become indicators for future monitoring. This foresight allows for timely interventions where resources or configurations can be adjusted before they lead to system degradation.

This same understanding informs the creation of preventive measures. By recognizing the early signs of anti-patterns, organizations can enforce best practices and integrate checks into their processes, particularly in areas such as deployment and configuration. These measures are tailored to the organization's specific Kubernetes environment, addressing unique challenges and avoiding generalizations.

Automating responses based on this knowledge becomes a strategic asset. Instead of broad-stroke automation, responses are nuanced, targeting specific aspects identified as problematic in past experiences. Therefore, automation becomes a dynamic tool, adapting to the evolving needs of the Kubernetes environment and continuously optimizing performance and stability.

Focused training and development stem from this understanding. Teams are trained not just in Kubernetes operations but in identifying and circumventing potential pitfalls. This targeted training approach ensures that teams are not only technically proficient but also adept at navigating the complexities of Kubernetes environments.

The practice of continuously reassessing and learning from past Kubernetes deployments fosters an environment of growth and adaptation. Teams don't just resolve current issues; they build resilience against future challenges, ensuring that their Kubernetes operations are not only effective in the present but prepared for the future.

By harnessing a thorough understanding of Kubernetes anti-patterns, organizations can move from reactive problem-solving to a proactive stance, enhancing their Kubernetes operations' efficiency, stability, and adaptability.

## Cultivating informed decision-making processes

When organizations recognize and comprehend the intricacies of Kubernetes anti-patterns, they are better positioned to make decisions that steer clear of these common pitfalls. This awareness becomes a guiding force in all aspects of Kubernetes management, from initial setup and configuration to ongoing maintenance and scaling.

The process begins with planning and strategy formulation. Armed with insights about what can go wrong and why, teams can plan their Kubernetes deployments more effectively. Decisions regarding architecture, resource allocation, and service configurations are made with a keen awareness of potential issues, leading to choices that are not only optimal for the current state but resilient for future demands.

Resource management decisions, a critical aspect of Kubernetes, are greatly influenced by this approach. Teams become adept at not only allocating resources efficiently but also at foreseeing scenarios where adjustments will be necessary. This preemptive thinking helps in avoiding scenarios where resources become either a bottleneck or are underutilized, ensuring balanced and cost-effective operations.

The choice of tools and technologies to complement Kubernetes operations is another area where informed decision-making plays a crucial role. Instead of being swayed by trends or vendor preferences, decisions are based on a clear understanding of how different tools interact with Kubernetes and the potential for anti-patterns. This leads to a more strategic selection of tools that align with the organization's specific needs and goals.

Security practices within Kubernetes environments also benefit from this informed approach. Understanding the security implications of certain configurations and the risks associated with common anti-patterns allows organizations to implement more robust security measures. Decisions surrounding access controls, network policies, and data encryption are made with a comprehensive view of potential vulnerabilities.

This influences how organizations respond to and learn from incidents. Post-incident reviews are not just about fixing the issue at hand but about dissecting the situation to understand what went wrong and why. These learnings are then fed back into the decision-making process, continually refining and improving Kubernetes practices.

## Guiding strategic planning and long-term vision

When organizations take into account the lessons learned from Kubernetes anti-patterns, their approach to technology architecture becomes more nuanced and forward-thinking. They are better equipped to design Kubernetes frameworks that are not only resilient against current operational challenges but also flexible enough to adapt to future technological shifts. This foresight in architectural decisions helps in avoiding rigid structures and overly complex configurations that might hamper scalability and adaptability down the line.

Resource management, a critical component of Kubernetes strategy, is profoundly influenced by the awareness of anti-patterns. Strategic planning that incorporates this knowledge leads to more effective and efficient resource utilization. Organizations develop a keen sense of balancing resources to avoid both over-provisioning, which can be costly, and underutilization, which can lead to performance bottlenecks. This balanced approach is vital not just for current efficiency but also for ensuring cost-effective growth and expansion in the future.

The rapid evolution of technology presents constant challenges and opportunities. A strategic vision informed by an understanding of Kubernetes anti-patterns equips organizations to remain agile and responsive to these changes. They can quickly adapt their Kubernetes strategies to new technologies and industry trends, maintaining relevance and effectiveness in a dynamic tech landscape.

## Promoting sustainable and scalable Kubernetes practices

The essence of sustainability in Kubernetes practices is rooted in resource efficiency. An informed approach, mindful of past missteps, guides organizations in optimizing resource use. This means creating Kubernetes environments that are precisely calibrated to use just the right amount of resources – not so much that it leads to wastage, and not so little that it impedes performance. Such efficiency is critical not only for operational cost savings but also for aligning with environmental sustainability goals.

Scalability is another pillar of sustainable Kubernetes practices. By understanding how previous configurations may have limited or hindered scalability, organizations can design systems that are inherently more flexible. This flexibility allows Kubernetes environments to expand or contract resource allocation seamlessly, accommodating fluctuating demands without necessitating a complete overhaul of the system. Scalability, in this sense, is not just a technical feature but a strategic approach that ensures Kubernetes can support the organization's growth and changing needs over time.

A key factor in achieving sustainable and scalable Kubernetes practices is integrating automation. Automation, tailored by insights from anti-patterns, becomes a tool for maintaining system health and efficiency. It involves automating routine tasks, sure, but it goes beyond that to include the automation of performance optimizations, resource scaling, and even some aspects of security management. This level of automation ensures consistency in operations and frees up valuable resources to focus on strategic initiatives rather than maintenance.

Another aspect is the continuous monitoring and refinement of Kubernetes environments. Promoting sustainability and scalability means regularly evaluating how the system is performing and making adjustments as needed. This ongoing process allows for the early detection of potential issues and the opportunity to optimize configurations before they become problematic. It's an approach that keeps the Kubernetes environment in a state of continual improvement, ensuring it remains efficient, effective, and aligned with the organization's evolving objectives.

## Improving organizational resilience to future challenges

Resilience in Kubernetes is, first and foremost, about building systems that can withstand and quickly recover from disruptions. This resilience is cultivated through a deep understanding of Kubernetes anti-patterns, which often highlight vulnerabilities and potential points of failure in the system. By identifying these areas, organizations can implement strategies such as robust failover mechanisms, effective disaster recovery plans, and comprehensive backup solutions. These strategies ensure that even in the face of unexpected failures or disruptions, the Kubernetes environment remains stable and recoverable.

Another key aspect of resilience is the ability to adapt to change, both in terms of technological advancements and evolving business requirements. Organizations improve their resilience by staying informed about the latest developments in Kubernetes and related technologies. This continuous learning enables them to adapt their Kubernetes strategies and practices to leverage new features and improvements, keeping their systems at the forefront of technological efficiency.

Resilience also involves fostering a culture of agility within the organization. Kubernetes teams should be encouraged to experiment, learn from their experiences, and continuously refine their skills and practices. This culture of agility and continuous improvement means that the organization is always ready to respond to new challenges, experiment with new solutions, and adapt its Kubernetes environment to meet changing demands.

Effective risk management is integral to organizational resilience. This involves not only identifying and mitigating risks associated with Kubernetes deployments but also having plans in place for dealing with potential future risks. Organizations can conduct regular risk assessments, stay alert to new security threats, and update their practices in line with best security practices to protect their Kubernetes environments from potential vulnerabilities.

Lastly, improving resilience is about developing a deep understanding of the organization's unique operational context and how Kubernetes fits within it. This understanding helps in tailoring the Kubernetes strategy to align with the organization's long-term goals and operational realities. It ensures that the Kubernetes environment is not just robust in a general sense but is specifically designed to support the unique needs and challenges of the organization.

By focusing on these aspects, organizations can significantly improve their resilience to future challenges in their Kubernetes environments. This resilience ensures that they are not only prepared to handle current operational demands but are also well-equipped to adapt and succeed in the face of future changes and challenges.

## Summary

This chapter focused on the root causes of Kubernetes anti-patterns and their impacts on system operations. It explored the historical evolution of Kubernetes, addressing misconceptions and knowledge gaps, and the role of architectural and organizational factors in Kubernetes deployments. This chapter also emphasized the human aspect of Kubernetes management, including the importance of skills, training, and communication.

Then, it examined how tooling and technology choices affect Kubernetes operations and highlighted the impact of anti-patterns on operational efficiency, such as altering development culture, disrupting workflows, and impacting service reliability. This chapter also discussed the importance of understanding these causes for developing predictive and preventive strategies and fostering a culture of continuous improvement. This chapter concluded by stressing the significance of a thorough understanding of Kubernetes anti-patterns in maintaining efficient, effective, and resilient Kubernetes environments.

In the next chapter, we will explore practical strategies for overcoming Kubernetes anti-patterns and implementing best practices, alongside methods for enhancing the Kubernetes environment through optimization techniques, advanced monitoring, and integrating cutting-edge technologies for a more efficient, secure, and resilient infrastructure.

# Part 2:
# Implementing
# Best Practices

In this part, you will acquire practical solutions, best practices, and insights from real-world case studies to effectively address Kubernetes anti-patterns throughout the Kubernetes ecosystem.

This part contains the following chapters:

# 4

# Practical Solutions
# and Best Practices

This chapter offers concise yet comprehensive guidance that targets the mitigation of Kubernetes anti-patterns through a series of effective strategies and recognized best practices. It directly addresses common issues such as suboptimal resource use, misconfigurations, and operational inefficiencies, offering practical solutions for each.

The chapter emphasizes the importance of making sound architectural decisions, implementing robust monitoring, and efficiently managing clusters to prevent these anti-patterns. Additionally, it highlights the critical role of skills development and clear communication among Kubernetes practitioners. This guide is designed not only to solve existing challenges but also to proactively enhance Kubernetes environments, making them more efficient, stable, and resilient against future operational complexities.

In this chapter, we will cover the following topics:

- Strategies to mitigate Kubernetes anti-patterns

- Implementing proven best practices

- Enhancing the Kubernetes environment

## Strategies to mitigate Kubernetes anti-patterns

To mitigate these anti-patterns, organizations need a comprehensive approach that encompasses various facets of Kubernetes deployment and management. This includes gaining insights into the underlying causes of these issues, which may stem from factors such as outdated practices, suboptimal configurations, or a lack of alignment with best practices as Kubernetes evolves.

Mitigation strategies also involve a deeper understanding of how Kubernetes impacts existing workflows and organizational dynamics. Successful navigation of Kubernetes anti-patterns requires a combination of technical expertise, effective tool choices, and alignment with organizational goals and culture.

In this exploration, we will delve into intricate factors contributing to Kubernetes anti-patterns and offer actionable strategies to address them.

## Customized solutions for diverse Kubernetes environments

Customized solutions for diverse Kubernetes environments involve a detailed and nuanced approach that considers the unique characteristics and requirements of each environment. This process is fundamental for effectively mitigating Kubernetes anti-patterns, as each deployment may present distinct challenges and demands.

The first, and perhaps most critical, step in crafting customized solutions is a deep dive into understanding the specifics of the Kubernetes environment. This understanding spans several dimensions: the scale of the deployment, the nature of the applications being run, the existing network infrastructure, security requirements, and the overarching organizational goals. For instance, a Kubernetes environment deployed for a large-scale, globally distributed application demands different considerations compared to a smaller, localized deployment. Understanding these nuances is key to identifying the right solutions.

With a clear understanding of the environment, the focus shifts to identifying the specific anti-patterns prevalent in that setup. In larger environments, common issues might include mismanaged resource allocation leading to cost inefficiencies or poorly implemented scaling strategies resulting in performance bottlenecks. In contrast, smaller environments might suffer from over-engineering or unnecessary complexities that hinder agility. Recognizing these patterns is essential for addressing them effectively.

Once specific anti-patterns are identified, the development of tailored strategies is the next critical step. This may involve a wide range of solutions, such as fine-tuning resource allocation to optimize costs and performance, revising network policies to enhance security and connectivity, or even restructuring the Kubernetes architecture to better suit the workload requirements. For example, transitioning to a microservices architecture might be beneficial for some environments, while others might benefit more from a serverless approach.

An important aspect of customizing solutions is ensuring that they integrate well with existing tools and operational workflows. This means that any solution should not only solve the immediate issues but also fit seamlessly into the organization's **continuous integration and deployment** (**CI/CD**) pipelines, monitoring systems, and other operational processes. This integration is crucial for maintaining a smooth and efficient workflow, minimizing disruption, and ensuring long-term sustainability.

Here are some examples of such solutions, tailored to address different scenarios in Kubernetes deployments:

- **For high-traffic applications**: In environments where Kubernetes is used to manage applications with high traffic volumes, customized solutions often focus on ensuring scalability and performance. An example would be implementing an advanced autoscaling strategy. This strategy could involve using **Horizontal Pod Autoscalers** (**HPAs**) in conjunction with Cluster Autoscalers. HPAs adjust the number of pods based on the current traffic and resource utilization, while Cluster Autoscalers manage the number of nodes in the cluster. This dual scaling mechanism ensures that the application can handle traffic spikes efficiently without overutilizing resources.

- **For security-centric deployments**: In environments where security is a paramount concern, such as in financial services or healthcare, a customized solution might involve implementing enhanced network policies and strict access controls. Utilizing Kubernetes network policies to control communication between pods and implementing a service mesh such as Istio can provide fine-grained control over network traffic. Additionally, integrating robust **identity and access management** (**IAM**) solutions, such as OAuth2 and **OpenID Connect** (**OIDC**), with Kubernetes **role-based access control** (**RBAC**) ensures that only authorized users and services can access sensitive resources.

- **For multi-cloud environments**: Organizations using Kubernetes across multiple cloud providers face unique challenges in maintaining consistency and optimizing costs. A customized solution here could involve implementing a unified deployment strategy using tools such as Terraform or Crossplane, which allow for declarative configuration of resources across different clouds. This approach simplifies management and ensures consistency across environments. Additionally, integrating cost-monitoring tools designed for multi-cloud environments can help in tracking and optimizing resource utilization and expenses.

- **For data-intensive workloads**: In environments with data-intensive applications, such as big data processing or **machine learning** (**ML**) workflows, customized solutions might focus on optimizing storage and data processing capabilities. This could include integrating Kubernetes with high-performance storage solutions such as Ceph or Portworx, which offer scalable and resilient storage options. Implementing StatefulSets in Kubernetes ensures that data-heavy applications maintain their state across pod restarts. Furthermore, setting up efficient data processing pipelines using Kubernetes Operators for specific databases or data processing frameworks can automate and optimize these workflows.

- **For small-scale or development environments**: In smaller-scale environments or development setups, the focus might be on simplicity and cost-effectiveness. A customized solution here could involve setting up a lightweight Kubernetes deployment using solutions such as Minikube or K3s, which are optimized for limited resources and simplicity. Additionally, integrating a simple CI/CD pipeline using tools such as Jenkins or GitLab CI can streamline the development and deployment process, making it easier for smaller teams to manage their Kubernetes deployments effectively.

- **For edge computing scenarios**: In edge computing environments, where resources are often constrained and latency is a critical factor, customized solutions could involve using lightweight Kubernetes distributions such as K3s, which are designed for resource-constrained environments. Additionally, implementing localized data processing and caching strategies, possibly using edge-optimized databases and storage solutions, can reduce latency and bandwidth requirements.

Each of these examples demonstrates how solutions in Kubernetes environments can be tailored to meet the specific requirements of different scenarios. By customizing strategies based on the unique needs of each deployment, organizations can optimize their Kubernetes environments for performance, security, cost-efficiency, and scalability.

## Streamlining DevOps processes to avoid pitfalls

Streamlining DevOps processes to avoid pitfalls involves specific actions and methodologies aimed at improving efficiency, reliability, and consistency in Kubernetes environments.

Here are some specifics on how organizations can customize their DevOps processes:

- **Automated CI/CD pipelines**: Implementing fully automated CI/CD pipelines is a cornerstone of streamlined DevOps in Kubernetes. Automation ensures consistent and error-free deployments. Tools such as Jenkins, GitLab CI, and Argo CD can be used to automate the deployment process. For example, Argo CD integrates with Kubernetes, allowing for automatic deployment and synchronization of applications based on Git repositories.

- **Infrastructure as Code (IaC)**: Using IaC tools such as Terraform or Ansible for provisioning and managing Kubernetes infrastructure ensures consistency and reduces manual errors. IaC allows DevOps teams to define and manage Kubernetes clusters and their associated resources using code, making it easier to implement changes, replicate environments, and roll back if needed.

- **GitOps for configuration management**: Adopting a GitOps approach for managing Kubernetes configurations can streamline the deployment process. In GitOps, the Git repository serves as the **single source of truth** (**SSOT**) for system configuration, ensuring that changes are traceable and reversible. This approach not only simplifies the management of Kubernetes configurations but also enhances collaboration and visibility across teams.

- **Container image management**: Streamlining the process of building, storing, and managing container images is vital. Implementing a robust container registry, such as Harbor or Docker Hub, and setting up automated image scanning for vulnerabilities ensures that only secure and compliant images are deployed to Kubernetes.

- **Monitoring and logging**: Integrating comprehensive monitoring and logging solutions into the DevOps pipeline is essential for the early detection of issues and performance optimization. Tools such as Prometheus for monitoring and Elasticsearch with Kibana for logging provide real-time insights into the Kubernetes environment, enabling quick identification and resolution of potential issues.

- **Automated testing**: Incorporating automated testing in the CI/CD pipeline is crucial for ensuring the reliability of applications. This includes unit tests, integration tests, and **end-to-end** (**E2E**) tests. Kubernetes-native testing frameworks such as Testcontainers or Sonobuoy can be used for this purpose, providing an environment that closely mimics production.

- **Feedback loops and continuous improvement**: Establishing feedback loops within the DevOps process allows for continuous improvement. This involves regularly reviewing and analyzing deployment practices, performance metrics, and incident reports to identify areas for improvement. Implementing tools for continuous feedback, such as Slack integrations for alerts, ensures that the team stays informed and can quickly respond to issues.

- **Simplifying rollbacks**: Ensuring the ability to quickly and easily roll back deployments in the event of a failure is critical. This can be facilitated through automated rollback mechanisms within the CI/CD pipeline, allowing teams to revert to the last stable version with minimal downtime.

## Implementing effective communication channels

Effective communication channels are vital for mitigating Kubernetes anti-patterns. Establishing a system where updates about deployments, configuration changes, and Kubernetes updates are communicated clearly and promptly is the first step. Integrating tools such as Slack or Microsoft Teams with Kubernetes environments can automate such updates, ensuring everyone stays informed in real time.

Creating a dedicated platform for technical discussions is essential. This could be a specialized forum or chat group where team members can discuss Kubernetes-specific issues, share insights, and collaboratively troubleshoot problems. This platform not only facilitates knowledge sharing but also aids in resolving issues before they escalate into larger problems.

Regular stakeholder meetings are crucial in maintaining a holistic view of the Kubernetes environment. These meetings, involving development, operations, and management teams, should focus on reviewing the current state of the Kubernetes infrastructure, addressing challenges, and planning future changes. This regular synchronization ensures that potential anti-patterns are identified and addressed collaboratively.

Maintaining comprehensive and accessible documentation is another key aspect. This includes detailed architecture descriptions, configuration guides, update logs, and troubleshooting manuals. Up-to-date documentation reduces misunderstandings and errors stemming from a lack of information or reliance on outdated practices.

Channels for feedback and suggestions encourage continuous improvement. Regular surveys, suggestion boxes, or open forums where team members can voice their feedback about the Kubernetes environment can reveal valuable insights into improvements or unidentified issues.

Breaking down silos between different teams to encourage cross-functional communication is important in a complex environment such as Kubernetes. This approach ensures a more holistic management of the Kubernetes environment, avoiding tunnel vision and ensuring that diverse perspectives contribute to the overall effectiveness of the deployment.

Remember – the implementation of effective communication channels in Kubernetes environments is a multi-dimensional strategy. It involves real-time updates, dedicated spaces for technical discussions, regular cross-team meetings, comprehensive documentation, open feedback mechanisms, and cross-functional collaboration. This comprehensive communication strategy is instrumental in mitigating Kubernetes anti-patterns, ensuring a well-informed, aligned, and collaborative approach to managing Kubernetes deployments.

The following tabular guide suggests how these practices might be adopted based on the size of the organization (small, medium, or large):

| Practice | Small Organizations | Medium Organizations | Large Organizations |
|---|---|---|---|
| Real-time updates | Use free or basic versions of tools such as Slack. | Invest in enterprise versions for better integration. | Utilize custom integrations and enterprise solutions. |
| Dedicated discussion platforms | Utilize open source forums or basic chat tools. | Set up specialized forums with more features. | Use enterprise-grade solutions with extensive support. |
| Regular meetings | Monthly or as-needed meetings. | Bi-weekly sprint reviews. | Weekly cross-departmental meetings. |
| Documentation | Maintain essential documentation on cloud services. | Develop comprehensive guides and update logs. | Implement a full documentation system with access control. |
| Feedback mechanisms | Simple online forms or direct emails. | Structured surveys and regular feedback sessions. | Comprehensive feedback systems with analytics. |
| Cross-functional communication | Occasional joint meetings with all staff. | Regular inter-departmental projects and meetings. | Structured cross-functional teams and leadership groups. |

## Role-based training and skills development

Role-based training and skills development are critical components in the strategy to mitigate Kubernetes anti-patterns. By tailoring training programs to the specific roles within a Kubernetes team, organizations can ensure that each team member possesses the necessary skills and knowledge to effectively manage and operate within the Kubernetes environment.

For developers, training focuses on best practices in containerization, efficient use of Kubernetes objects such as pods, services, and deployments, and understanding how to design applications that are Kubernetes-friendly. This involves not just technical know-how but also an appreciation of the Kubernetes philosophy and how it impacts application architecture.

Operations teams require a different set of skills. Their training emphasizes Kubernetes cluster management, monitoring, troubleshooting, and ensuring **high availability** (**HA**). Operations personnel need to be adept at using tools such as Prometheus for monitoring, fluent in navigating the Kubernetes Dashboard, and proficient in implementing **disaster recovery** (**DR**) strategies.

For security personnel, Kubernetes training includes understanding network policies, managing RBAC, securing container images, and integrating security at every level of the Kubernetes stack. This is crucial in an era where security is paramount, and Kubernetes environments are often targeted due to their critical role in the infrastructure.

**Quality assurance** (**QA**) professionals also benefit from Kubernetes-specific training. Their focus is on understanding how Kubernetes affects testing strategies, setting up effective testing environments within Kubernetes, and ensuring that applications perform reliably in a Kubernetes context.

Customizing these training programs to fit the needs of each role ensures that the entire team is not only proficient in their respective areas but also understands how their role fits into the larger Kubernetes ecosystem. This holistic understanding is key in preventing siloed working methods, which can often lead to anti-patterns and inefficiencies.

In addition to formal training, creating opportunities for hands-on experience is vital. This can be achieved through internal workshops, hackathons, or allowing team members to rotate through different roles within the Kubernetes environment. Such experiences encourage a deeper understanding of Kubernetes and foster a culture of continuous learning.

Encouraging certification in Kubernetes, such as the **Certified Kubernetes Administrator** (**CKA**) or **Certified Kubernetes Application Developer** (**CKAD**), is another effective way to ensure that team members possess a standardized level of knowledge and skill.

Moreover, providing access to ongoing learning resources such as online courses, webinars, and attendance at industry conferences keeps the team updated with the latest developments in Kubernetes and related technologies.

The following is a sample role-based training matrix for a Kubernetes environment. This matrix outlines key roles involved in Kubernetes operations and the recommended areas of training for each role:

| Role | Core Training Areas | Additional Skills |
| --- | --- | --- |
| Developers | <ul><li>Kubernetes basics</li><li>Containerization with Docker</li><li>Designing Kubernetes-friendly applications</li><li>Using Kubernetes API</li><li>Implementing CI/CD pipelines</li></ul> | <ul><li>Microservices architecture</li><li>Serverless on Kubernetes</li><li>Application performance optimization</li></ul> |
| Operations team | <ul><li>Kubernetes cluster management</li><li>Monitoring and logging</li><li>Network configuration</li><li>DR and backup strategies</li><li>Security best practices</li></ul> | <ul><li>Automation and scripting</li><li>Cloud provider-specific Kubernetes services</li><li>Advanced troubleshooting techniques</li></ul> |

| Role | Core Training Areas | Additional Skills |
|---|---|---|
| Security personnel | • Kubernetes network policies<br>• RBAC<br>• Securing container images<br>• Integrating security tools with Kubernetes<br>• Security auditing and compliance | • Vulnerability assessment<br>• Security in DevOps (DevSecOps)<br>• Encryption and data protection techniques |
| QA | • Testing strategies in Kubernetes<br>• Setting up Kubernetes testing environments<br>• Performance and load testing<br>• Automated testing frameworks | • Chaos engineering<br>• User experience testing in containerized applications<br>• Continuous testing in CI/CD pipelines |
| DevOps engineers | • Implementing Kubernetes in DevOps workflows<br>• CI/CD tools<br>• IaC<br>• Kubernetes scalability and optimization<br>• Cross-functional collaboration techniques | • Cloud-native development practices<br>• Advanced CI/CD techniques<br>• Observability and analysis in Kubernetes |

## Structuring teams for efficient Kubernetes management

Structuring teams for efficient Kubernetes management calls for a thoughtful approach that aligns with the complexities and dynamic nature of Kubernetes. The focus is on crafting teams that are versatile, well informed, and highly collaborative.

At the core of this structure are cross-functional teams. These teams combine diverse expertise, drawing from development, operations, and security. Developers in these teams are not just code focused; they need to understand how their applications will be deployed and managed in Kubernetes. They work closely with operations experts, who bring in-depth knowledge of managing Kubernetes clusters, ensuring smooth deployments and handling the intricacies of cluster management. Security experts in the team are responsible for embedding security practices into the deployment pipeline, safeguarding applications right from their development stages.

The composition of these teams reflects the diversity of tasks in Kubernetes management. It's not just about having experts in individual fields; it's about fostering a culture where these experts collaborate seamlessly. For instance, in deploying a new application, a developer, operations specialist, and security expert would work in tandem to ensure that the application is not only functionally sound but also optimally configured and secure.

A critical aspect of this team structure is flexibility in roles. While each member has their primary area of expertise, they are encouraged to have a working understanding of other aspects of Kubernetes. This cross-training ensures that the team can pivot quickly in response to various challenges. For example, when a security expert understands the basics of application development, they can foresee potential security issues earlier in the development cycle.

Defining clear roles and responsibilities is essential in avoiding overlaps and ensuring that every critical aspect of Kubernetes management is attended to. This clarity is about knowing who is responsible for which part of the Kubernetes ecosystem, from deploying applications to monitoring and maintaining cluster health. Such a defined structure brings a sense of accountability and order, critical in managing complex systems such as Kubernetes.

Training and development are embedded into the team's routine. Given the ever-evolving nature of Kubernetes, staying updated with the latest features, best practices, and emerging trends is non-negotiable. Regular training sessions, whether through external courses or internal workshops, are scheduled. Knowledge sharing is encouraged, with team members presenting insights from recent projects or learnings from external training. This continuous learning approach ensures that the team remains adept and agile in handling the Kubernetes environment.

A psychologically safe work environment complements this structural approach. In such an environment, team members feel comfortable sharing ideas, discussing challenges openly, and learning from mistakes. This aspect is crucial in a field that is as fast-paced and complex as Kubernetes management. It fosters an atmosphere where innovative problem-solving thrives, and continuous improvement is a collective goal.

Regular strategy sessions are a staple. These sessions provide a forum for teams to review their workflows, discuss challenges faced, brainstorm solutions, and plan for future projects. It's a time for reflection and proactive planning, where the team's structure and processes are reassessed and realigned with the evolving demands of Kubernetes management.

Communication within and between these teams is streamlined. Regular meetings, clear documentation of processes and decisions, and established communication protocols ensure that everyone stays on the same page. This streamlined communication is vital in a domain where a small miscommunication can lead to significant issues in the Kubernetes environment.

In crafting teams for effective Kubernetes management, the focus is on creating a balance between individual expertise and collaborative synergy. It's about structuring teams in a way that they are greater than the sum of their parts, capable of navigating the complexities of Kubernetes with competence and confidence. This approach not only ensures efficient management of Kubernetes environments but also contributes to the professional growth and satisfaction of each team member.

## Embracing Agile methodologies in Kubernetes projects

Implementing Agile methodologies in Kubernetes projects transforms the management and deployment of these systems. It begins with the adoption of iterative development cycles or sprints, which break down complex Kubernetes tasks into manageable segments. This approach, crucial in handling the inherent complexities of Kubernetes, allows teams to focus on specific areas such as updating clusters, enhancing security, or optimizing resource allocation in distinct phases. Each phase or sprint brings its own set of goals, deliverables, and timelines, making the process more organized and manageable.

Regular feedback loops and sprint reviews are integral to this Agile integration. After each development cycle, the team assesses the work against predetermined objectives. This step is more than just a progress check; it's an opportunity to gather valuable feedback, identify areas for improvement, and refine strategies for subsequent sprints. In Kubernetes projects, these adjustments might involve reconfiguring resources, updating automation scripts, or modifying security protocols based on real-time feedback and observations.

Collaboration takes a front seat in Agile methodologies. Cross-functional teams comprising developers, operations staff, and QA professionals collaborate closely, ensuring that Kubernetes deployments are not only developed efficiently but are also seamlessly integrated into existing systems and workflows. This collaborative approach is essential in Kubernetes environments, where the success of the application depends on how well it's integrated and managed within the Kubernetes ecosystem.

A user-centric focus is another hallmark of Agile methodologies. By continuously releasing and updating features and gathering user feedback, teams can ensure that their Kubernetes applications meet user needs and expectations more accurately. This approach might involve using Kubernetes features such as canary deployments, where new application versions are gradually rolled out to a subset of users, allowing teams to gather user feedback and make adjustments before a full-scale launch.

Daily stand-up meetings keep the team aligned and informed. In these brief, focused meetings, team members discuss their progress and any obstacles they are facing. Given the dynamic nature of Kubernetes, where changes are frequent and rapid, these daily meetings are crucial in maintaining project momentum and addressing issues promptly.

Simplicity and sustainability in system design and processes are encouraged in Agile. This means creating Kubernetes configurations and workflows that are as straightforward as possible, minimizing complexity, and automating routine tasks to improve efficiency and reduce errors.

Flexibility and adaptability are key components of Agile methodologies. Teams are encouraged to remain open to changes and adapt their strategies in response to evolving project requirements, technological advancements, and changing business landscapes. This flexibility is especially important in Kubernetes environments, which are subject to continuous change and evolution.

Incorporating Agile methodologies in Kubernetes projects, therefore, is not just about applying a set of principles; it's about creating a dynamic, responsive, and collaborative environment. This environment is conducive to managing the complexities of Kubernetes, ensuring that projects are not only technically sound but also aligned with user needs and business objectives.

## Establishing robust Kubernetes governance policies

Establishing robust Kubernetes governance policies involves creating a comprehensive set of rules and guidelines that dictate the use and management of Kubernetes within an organization. These policies cover a wide range of areas, including security, compliance, resource management, and operational best practices.

Developing clear standards for cluster setup and management is the foundation of Kubernetes governance. This includes policies on networking, storage, and compute configurations. For example, detailed guidelines on network policies are essential to ensure the isolation and security of different applications within the cluster. Additionally, setting standards for resource quotas and limits is crucial to prevent resource hogging and ensure fair usage across different teams or applications.

Security is at the forefront of Kubernetes governance. Access control policies, especially those implementing RBAC, are essential to ensure that users have only the permissions necessary for their role. Policies around container image security are equally important, often mandating the use of image-scanning tools to detect vulnerabilities. Secure management of secrets and sensitive data is another area where strict governance is necessary, with policies often dictating the use of Kubernetes Secrets or external secrets management systems.

Compliance with regulatory standards is another critical aspect. Kubernetes governance policies must ensure that the organization's use of Kubernetes adheres to relevant data privacy laws, financial regulations, and industry-specific standards. This involves setting policies for data encryption, logging, and ensuring data residency.

Operational efficiency is enhanced through governance policies that establish best practices for deploying applications, managing resources, and handling service disruptions. For example, requiring all deployments to pass through a CI/CD pipeline and incorporating automated testing can significantly reduce the risk of deployment-related issues.

Monitoring and **incident response** (**IR**) are also governed by specific policies. Organizations often define which metrics and logs should be collected, how monitoring should be performed, and the procedures for responding to incidents. Tools such as Prometheus for monitoring and the ELK Stack for log management are commonly specified in these governance policies.

To give a clearer picture, here is a sample table of Kubernetes governance policies:

| Policy Area | Sample Policy |
|---|---|
| Cluster configuration | All clusters must be configured with network policies to isolate namespaces. |
| Access control | Implement RBAC with the principle of least privilege (PoLP). All user access must be reviewed quarterly. |
| Container security | All container images must be scanned for vulnerabilities before deployment. |
| Secrets management | Use Kubernetes Secrets for managing sensitive data, with encryption at rest and in transit. |
| Compliance | Ensure logging and monitoring practices comply with the General Data Protection Regulation (GDPR) for handling user data. |
| Resource management | Set resource quotas for namespaces to prevent overconsumption by a single team or application. |
| Deployment practices | All application deployments must go through automated CI/CD pipelines with required testing stages. |
| Monitoring and reporting | Use Prometheus for monitoring cluster performance and set up alerts for critical thresholds. |
| IR | Establish an IR protocol, including immediate notification and post-incident analysis. |
| Regular policy review | Review and update governance policies bi-annually or as major Kubernetes updates are released. |

## Advanced error tracking and reporting mechanisms

Advanced error tracking and reporting mechanisms in Kubernetes environments are integral for maintaining robust and reliable systems. These mechanisms involve a combination of sophisticated tools and methodologies designed to capture, analyze, and respond to errors in real time.

Central to this setup is the integration of powerful logging tools such as Elasticsearch, Fluentd, and Kibana, commonly referred to as the EFK stack. Elasticsearch acts as a search and analytics engine, storing and indexing logs for easy retrieval. Fluentd collects logs from various sources in the Kubernetes cluster, including nodes and pods, and feeds them into Elasticsearch. Kibana then provides a user-friendly interface for querying the logs and visualizing the data. This setup enables teams to quickly sift through massive amounts of log data to identify and understand the root causes of errors.

**Application performance monitoring** (**APM**) tools such as New Relic, Datadog, or Dynatrace are also crucial. These tools provide insights into the performance of applications running in Kubernetes. They help in identifying performance anomalies, tracking response times, and understanding the impact of errors on application behavior. APM tools are particularly valuable because they offer granular visibility into the application, often pinpointing issues down to specific lines of code or API calls.

Alerting mechanisms form another crucial component. Tools such as Prometheus can be used to monitor a wide range of metrics from Kubernetes clusters. When integrated with alert managers, these tools can trigger notifications based on predefined criteria or detected anomalies. These alerts ensure that the relevant team members are promptly informed about issues, enabling quick response and resolution.

Distributed tracing is vital in diagnosing errors in microservices architectures common in Kubernetes. Tools such as Jaeger or Zipkin trace the flow of requests through the various services, providing a clear picture of where failures or performance issues occur. This level of tracing is indispensable in complex environments, where pinpointing the exact location of an issue can be challenging.

Beyond detection, advanced error tracking in Kubernetes often includes automating the response to certain types of errors. For example, Kubernetes might automatically scale up resources in response to a performance bottleneck or roll back a deployment if a critical error is detected. Automation not only speeds up the response to issues but also reduces the potential for human error.

Managing and analyzing logs effectively is another critical aspect. With the high volume of log data generated in Kubernetes environments, setting up policies for log retention and analysis is essential. Deciding which logs to keep, at what level of detail, and for how long are important considerations. Advanced log analysis techniques, such as ML algorithms, can be employed to sift through this data, identifying patterns and predicting potential issues before they become critical.

Creating comprehensive dashboards using tools such as Grafana is also part of advanced error tracking. These dashboards provide a visual overview of the health and performance of the Kubernetes environment. Customizable dashboards are particularly useful as they can be tailored to show relevant information for different roles, from developers needing detailed application insights to operations teams monitoring the overall health of the cluster.

Incorporating these advanced error tracking and reporting mechanisms in Kubernetes environments ensures not just the detection of issues but also their in-depth analysis and prompt resolution. This approach is crucial for maintaining the high reliability and performance standards expected in modern Kubernetes deployments.

## Integrating security from the development phase

Integrating security from the development phase in Kubernetes projects involves a holistic approach where security considerations are embedded into every aspect of the application lifecycle, right from the initial design. This approach, often termed *shifting security left*, is vital in creating a secure Kubernetes environment.

The integration begins at the planning and architectural design stage. Here, security is a primary consideration in the design of microservices, data flow, and component isolation within the Kubernetes cluster. Adopting principles such as least privilege and zero trust at this stage ensures that each application component operates with minimal permissions necessary for its function.

As the development progresses, incorporating code analysis tools is crucial. **Static application security testing (SAST)** and **dynamic application security testing (DAST)** tools are integrated into the development workflow. These tools proactively scan the code base for potential security vulnerabilities, such as insecure coding practices or known vulnerabilities in dependencies, enabling developers to rectify issues at an early stage.

Container security forms a core part of this approach. It involves scanning container images for vulnerabilities during the build process and continuously thereafter. Tools such as Clair and Trivy can be integrated into CI/CD pipelines for automated scanning, ensuring that container images are secure before deployment.

IAM in Kubernetes is also critical. Implementing RBAC effectively manages access to the Kubernetes API. Managing credentials and keys securely, and ensuring their regular rotation, are essential practices to maintain tight control and monitoring of access to Kubernetes resources.

Network security within Kubernetes necessitates early integration. Setting up network policies to control traffic flow between pods ensures that services are accessible only to necessary components. Tools such as Calico or Cilium enforce these policies, providing a layer of security against unauthorized access and lateral movements within the cluster.

Security considerations extend to the deployment process. Techniques such as rolling updates and canary deployments minimize risks during updates. The deployment process must be reversible to roll back changes in case of security issues. Continuous monitoring of the runtime environment for real-time detection and response to security incidents is a critical practice.

Education and awareness among the development team are equally important. Regular training sessions on secure coding practices, keeping the team updated about the latest security threats, and workshops on effective use of security tools cultivate a security-conscious culture within the team.

By embedding security into every stage of the application lifecycle in Kubernetes environments, organizations can significantly reduce the risk of vulnerabilities and enhance their security posture. This proactive approach to security ensures that Kubernetes deployments are not just functional and efficient but also secure by design.

In wrapping up our initial discussion, we've examined a broad spectrum of methods to address challenges commonly encountered in Kubernetes environments. This has covered everything from enhancing communication within teams to leveraging the collective wisdom of the Kubernetes community. Our goal has been to empower both individuals and teams with the knowledge and tools needed to improve their management and oversight of Kubernetes projects.

Looking ahead, we will transition from mitigating risks to actively enhancing our Kubernetes operations. We'll explore foundational design principles and strategic approaches for resource management, ensuring system resilience, and maximizing performance. By implementing these well-established practices, you'll be better equipped to optimize your Kubernetes setup, boosting both security and efficiency across your deployments.

# Implementing proven best practices

Implementing proven best practices in Kubernetes transcends mere operational efficiency; it is an essential pathway to mastering the platform's vast capabilities. This exploration delves into the refined and validated strategies that form the cornerstone of effective Kubernetes management. Covering a broad spectrum from architectural design principles to operational procedures, these best practices are the culmination of collective wisdom from the Kubernetes community. They serve as a guide to navigating the intricacies of Kubernetes, ensuring environments are not just robust and secure but also optimized for peak performance and scalability. Embracing these practices paves the way for mastering Kubernetes, turning its complexities into strategic advantages.

## Core principles of Kubernetes architecture design

Implementing proven best practices in Kubernetes architecture design revolves around several core principles. Each of these principles plays a critical role in shaping robust, scalable, and efficient Kubernetes environments.

Here's a detailed breakdown of these core principles:

- **Declarative configuration and automation**: In Kubernetes, the management of resources is done declaratively. Users define the desired state of their application or component within a configuration file. Kubernetes continually works to maintain this state, automating deployment and recovery processes. This approach reduces manual interventions, minimizes errors, and streamlines management.

- **Modularity and microservices architecture**: Kubernetes is ideally suited for a microservices architecture. It encourages breaking down applications into smaller, independent modules (microservices). This modularity enhances scalability, as each microservice can be scaled independently based on specific needs. It also facilitates easier updates and faster development cycles.

- **HA and fault tolerance (FT)**: Kubernetes architecture is built to support HA and FT. Features such as replication controllers and replica sets ensure applications are always running and accessible. If a pod fails, Kubernetes automatically replaces it, and if a node goes down, pods are rescheduled on healthy nodes. Designing stateless applications further reinforces this, as they are easier to manage and scale in distributed systems.

- **Efficient resource management**: Kubernetes offers sophisticated tools for managing computing resources such as CPU and memory. Administrators can set resource requests and limits for pods, ensuring optimal resource allocation. This approach prevents resource contention and maximizes infrastructure utilization, leading to better application performance.

- **Load balancing and service discovery**: Kubernetes provides in-built mechanisms for load balancing and service discovery. It automatically distributes network traffic to pods and offers stable endpoints for services through its service abstraction. This ensures that services are easily discoverable within the cluster and traffic is efficiently managed.

- **Inherent security measures**: Security in Kubernetes is not an afterthought but is integrated into its architecture. It involves setting up robust access controls such as RBAC, securing intra-cluster communication with TLS encryption, and ensuring container images are secure. Kubernetes' design encourages a security-first approach in all aspects of cluster management.

- **Observability**: Effective monitoring, logging, and tracing are fundamental in Kubernetes. These observability tools provide vital insights into the cluster's operations, helping administrators to quickly diagnose issues, understand application performance, and make informed decisions regarding scaling and resource allocation.

Each of these principles contributes to creating a Kubernetes environment that is not only tailored to current operational needs but is also prepared for future scalability and adaptability challenges. By adhering to these core principles, organizations can harness the full potential of Kubernetes, ensuring their deployments are robust, efficient, and secure.

## Effective load-balancing strategies

Effective load-balancing strategies are crucial in Kubernetes to ensure optimal distribution of network traffic and efficient resource utilization. Implementing these strategies involves several approaches, each tailored to manage traffic flow to applications running within a Kubernetes cluster.

Here's a detailed look at these strategies:

- **Service-based load balancing**: Kubernetes uses Services as an abstract way to expose applications running on a set of Pods. Services manage load balancing and provide a single point of entry for accessing Pods. This approach decouples the frontend exposure from backend workloads, ensuring that clients are not affected by changes in Pods.

- **Ingress Controllers and load balancers**: For external traffic, Kubernetes Ingress Controllers are used. They provide HTTP and HTTPS routing to services based on defined rules. Ingress resources are configured to manage external access to the services, often integrating with cloud provider load balancers or using internal load balancers for more control and customization.

- **NodePort and ClusterIP services**: Kubernetes offers NodePort and ClusterIP services for internal load balancing. NodePort exposes services on each Node's IP at a static port, allowing external traffic access through these node ports. ClusterIP, on the other hand, provides internal load balancing within the cluster, making services reachable within the cluster network.

- **HPA**: To dynamically handle varying loads, the HPA automatically scales the number of Pods in a deployment, replication controller, or replica set based on observed CPU utilization or other selected metrics. The HPA ensures that the load is spread across enough Pods to handle it efficiently.

- **Pod affinity and anti-affinity**: Kubernetes allows setting up pod affinity and anti-affinity rules. These rules control how Pods are grouped together or separated across different nodes in the cluster. By intelligently placing Pods based on the workload, you can enhance load balancing and improve resource utilization.

- **Network policies for traffic control**: Implementing network policies in Kubernetes can control how Pods communicate with each other and with other network endpoints. By defining appropriate network policies, you can direct traffic flow more effectively, ensuring that it is balanced and secure.

- **Session affinity**: For certain applications, it's crucial to maintain client session affinity (also known as sticky sessions). Kubernetes Services can be configured for session affinity, ensuring that all requests from a particular client are sent to the same Pod, as long as it is available.

- **Custom load-balancing algorithms**: Kubernetes allows the use of custom load-balancing algorithms through external or third-party load balancers. These can be tailored to specific application needs, such as least connections, IP hash, or custom hashing methods, providing more granular control over traffic distribution.

By implementing these effective load-balancing strategies, Kubernetes ensures that applications are not only highly available but also resilient to fluctuations in traffic, maintaining optimal performance and user experience. These strategies contribute significantly to the robustness and efficiency of applications running in Kubernetes environments.

## Implementing comprehensive backup and recovery plans

Implementing comprehensive backup and recovery plans in Kubernetes is crucial for ensuring data integrity and availability, particularly in the event of failures, data corruption, or other unforeseen incidents. A well-thought-out backup and recovery strategy encompasses various components of the Kubernetes environment, from the application data to the cluster state.

Let's break down backup and DR plans into two distinct sections and explore different types of DR strategies in Kubernetes environments.

### Backup plans in Kubernetes

- **Application data backup**: This involves regularly backing up the data of stateful applications running in Kubernetes. Tools such as Velero or Stash can be used to automate the backup of data stored in **Persistent Volumes** (**PVs**). The frequency and timing of backups should be based on data criticality and change rate.

- **Cluster configuration backup**: Backing up Kubernetes cluster configurations, including resource definitions (deployments, services, and so on), is essential. This ensures that you can quickly restore the cluster's operational state. Tools such as Velero can also capture and back up these configurations.

- **etcd database backup**: The `etcd` database is Kubernetes' primary data store. Regularly backing up `etcd` is crucial for recovering the cluster's state in case of corruption or loss. `etcdctl snapshot save` is commonly used for this purpose.

- **Automated and scheduled backups**: Automation of backup processes minimizes human error and ensures consistent data protection. Utilizing cron jobs or Kubernetes CronJobs to schedule backups can achieve this automation.

- **Offsite and redundant storage**: Backups should be stored offsite or replicated across multiple locations to protect against site-specific disasters. Cloud storage solutions are often used for their scalability and geographic distribution capabilities.

- **Backup data security**: Encrypting backup data and controlling access to it is as important as securing primary data. Implement strong encryption and access control policies for backup data.

- **Regular testing of backups**: Periodically test backup restoration processes to ensure data integrity and the effectiveness of the backup strategy.

- **Data retention policies**: Specify how long backups are kept before they are deleted. This ensures compliance with legal and regulatory requirements and optimizes storage usage. Setting clear retention rules helps manage the lifecycle of backup data systematically, preventing unnecessary storage consumption and maintaining a clean backup environment.

- **Automatic pruning of outdated backups**: Reduces storage costs and management overhead, ensuring that only relevant backups are retained. Implementing automatic pruning involves configuring backup tools to delete old backups at regular intervals, thus maintaining an efficient and cost-effective backup repository.

- **Incremental backup implementation**: Capture only changes made since the last backup, reducing the backup size and minimizing storage requirements to enhance backup efficiency and decrease the time needed for backups. Configuring backup systems to perform incremental rather than full backups can significantly optimize resource use and improve recovery times.

## DR strategies

- **Multi-zone/region availability**: Deploying Kubernetes clusters across multiple zones or regions provides resilience against zone-specific failures. If one zone goes down, the other can continue to operate, minimizing downtime.

- **Active-passive configuration**: In this strategy, one Kubernetes cluster is active (handling production traffic) while another is passive (on standby). The passive cluster can be brought online in case the active cluster fails. Regular synchronization and backup restoration are used to keep the passive cluster updated.

- **Active-active configuration**: Here, two or more clusters run simultaneously, handling production traffic. They are often geographically distributed. This setup provides HA as traffic can be rerouted to the other active cluster(s) in case of a failure.

- **Cloud-based DR solutions**: Utilizing cloud providers' DR solutions can offer added layers of resilience. These solutions often come with built-in tools for data replication, backup, and quick recovery.

- **On-premises-to-cloud DR**: For on-premises Kubernetes environments, replicating critical data and configurations to a cloud environment can provide an effective DR solution. In case of major on-premises failures, the cloud environment can take over.

- **Regular DR testing**: Conducting regular DR drills ensures that the **DR plan** (**DRP**) is effective and the team is prepared to execute it in case of an actual disaster.

## Kubernetes versioning and upgrade best practices

Effectively managing Kubernetes versioning and upgrades is crucial for maintaining a stable, secure, and efficient environment. Staying current with Kubernetes versions ensures access to the latest features, performance improvements, and security patches. Here's a detailed look at best practices for Kubernetes versioning and upgrade processes:

- **Understanding release channels and versioning scheme**: Kubernetes follows a versioning scheme that includes major, minor, and patch releases. Familiarize yourself with this scheme to understand what each upgrade entails. Major releases (1.x) might introduce significant changes, while minor (1.x.y) and patch releases (1.x.y.z) typically include bug fixes and minor improvements.

- **Staying informed on release notes**: Before planning an upgrade, review the release notes for the new version. These notes provide critical information on changes, deprecations, bug fixes, and known issues, which are essential for assessing the impact on your current environment.

- **Regularly scheduled upgrades**: Implement a regular schedule for reviewing and applying new releases. Staying up to date with recent versions helps avoid the pitfalls of outdated software, such as security vulnerabilities and compatibility issues.

- **Testing in a staging environment**: Before applying an upgrade to your production environment, test it in a staging environment that closely mirrors your production setup. This includes testing all applications, services, and integrations to ensure they work as expected with the new version.

- **Automated backup before upgrading**: Ensure that you have automated backups of critical components, such as cluster data, configurations, and application data. This step is crucial for recovery in case the upgrade introduces unexpected issues.

- **Phased rollout of upgrades**: For large and complex environments, consider a phased rollout of the upgrade. Start with less critical clusters or namespaces to gauge the impact before proceeding to more critical parts of your environment.

- **Utilizing canary deployments**: Canary deployments involve upgrading a small portion of your cluster first. This approach allows you to monitor the performance and stability of the new version before rolling it out to the entire cluster.

- **Monitoring post-upgrade**: After an upgrade, closely monitor the cluster for any anomalies. This includes checking system logs, application performance, and resource utilization to ensure everything is functioning as expected.

- **Rollback strategy**: Have a clear rollback strategy in case the upgrade doesn't go as planned. This should include steps to revert to the previous stable version without impacting running applications.

- **Compliance and compatibility checks**: Ensure that the new version complies with your organizational policies and maintains compatibility with existing tools and integrations.

## Securing Kubernetes Secrets management

Securing the management of Secrets in Kubernetes is a critical aspect of safeguarding sensitive data such as passwords, tokens, and keys within your Kubernetes environment. Effective Secrets management not only protects against unauthorized access but also ensures the integrity and confidentiality of the data throughout its lifecycle. Here's a comprehensive approach to securing Kubernetes Secrets management:

- **Understanding Kubernetes Secrets**: Begin by familiarizing yourself with the Kubernetes Secrets object. Secrets in Kubernetes are used to store and manage sensitive information, such as passwords, OAuth tokens, and SSH keys. Understanding how Secrets are used and accessed by Pods in Kubernetes is foundational to implementing effective security measures.

- **Encrypting Secrets at rest**: Ensure that Secrets are encrypted at rest within the `etcd` database. By default, Secrets are stored as plaintext in `etcd`; enabling encryption at rest is vital to prevent unauthorized access to sensitive data, especially in case of a breach or compromised `etcd` access.

- **Using namespaces wisely**: Leverage Kubernetes namespaces to limit the scope of Secrets. Namespaces can be used to isolate Secrets within specific areas of your cluster, reducing the risk of accidental exposure or unauthorized access from other parts of the cluster.

- **RBAC**: Implement RBAC to control which users and Pods have access to Secrets. RBAC policies should follow PoLP, ensuring that users and applications have only the permissions necessary for their function.

- **Audit logging and monitoring**: Enable audit logging to track access and changes to Secrets. Monitoring access logs helps in detecting unauthorized attempts to access Secrets and ensures compliance with auditing requirements.

- **Secrets rotation and expiry**: Regularly rotate Secrets and set expiry dates where applicable. Automated rotation of Secrets minimizes the risk associated with long-term exposure or compromise of a Secret.

- **Using external secrets management tools**: Consider integrating external secrets management systems such as HashiCorp Vault, AWS Secrets Manager, or Azure Key Vault. These systems offer advanced features for secrets management, such as dynamic secrets, fine-grained access policies, and automatic rotation.

- **Avoid hardcoding Secrets**: Never hardcode Secrets in application code or Docker images. Instead, use Kubernetes Secrets to inject sensitive data into Pods at runtime.

- **Secure Secret injection into Pods**: Use mechanisms such as environment variables or volume mounts to securely inject Secrets into Pods. When using environment variables, be cautious as they can be exposed to any process within the Pod and might appear in logs or error messages.

- **Regularly reviewing and auditing Secrets**: Conduct periodic audits of your Secrets to ensure that they are still in use, have the correct access policies, and comply with organizational security policies. Unused or orphaned Secrets should be removed to reduce the attack surface.

## Efficient log management and analysis

Efficient log management and analysis in Kubernetes are crucial for maintaining operational insight, troubleshooting issues, and ensuring compliance with auditing requirements. Given the distributed nature of Kubernetes, dealing with logs can be complex. Here's a detailed approach to efficiently managing and analyzing logs in Kubernetes:

- **Centralized logging**: Implement a centralized logging system to aggregate logs from all components of the Kubernetes cluster. This includes logs from the Kubernetes master, nodes, pods, and the applications running inside those pods. Centralized logging provides a holistic view of the cluster's state and behavior, crucial for effective troubleshooting and analysis.

- **Choosing the right tools**: Tools such as Elasticsearch, Fluentd, and Kibana (EFK stack) or a combination of Prometheus and Grafana are popular choices for Kubernetes logging. Elasticsearch acts as a powerful search and analytics engine, Fluentd collects logs from various sources, and Kibana provides user-friendly interfaces for querying and visualizing logs. Prometheus, coupled with Grafana, is excellent for monitoring and visualizing time-series data.

- **Structured logging**: Implement structured logging within applications. Structured logs are easier to query and analyze compared to plain text logs. They contain consistent and machine-readable data, typically in JSON format, which makes automated analysis and querying more straightforward.

- **Log rotation and retention policies**: Set up log rotation and define retention policies to manage the storage of logs efficiently. Log rotation prevents files from becoming too large, while retention policies ensure that logs are stored for an appropriate amount of time, balancing between operational needs and storage constraints.

- **Real-time monitoring and alerting**: Integrate real-time monitoring and alerting into your logging system. Tools such as Prometheus can be configured to trigger alerts based on specific log patterns or anomalies, enabling quick responses to potential issues.

- **Efficient storage management**: Logs can consume significant storage space. Utilize efficient storage solutions and consider compressing logs to reduce storage requirements. When using cloud services, take advantage of cloud storage options that offer scalability and cost-effectiveness.

- **Log analysis and visualization**: Employ log analysis tools and techniques to extract meaningful insights from log data. Visualization tools such as Grafana can be used to create dashboards that provide an at-a-glance view of log data, making it easier to spot trends, anomalies, or issues.

- **Security and access control**: Secure your log data and control access to it. Ensure that sensitive data in logs is encrypted and that access to logs is controlled using RBAC.

- **Compliance and auditing**: Ensure your log management strategy aligns with compliance requirements. This includes capturing all relevant log data, storing it securely, and making it available for auditing purposes.

- **Regular review and optimization**: Regularly review your log management and analysis practices. As your Kubernetes environment evolves, so too should your logging strategy to ensure it remains efficient and effective.

Having explored a suite of best practices to refine our Kubernetes operations, we've covered everything from architectural foundations to advanced management of Kubernetes APIs and security. These insights are aimed at not only preventing issues but also elevating the operational standards of your Kubernetes deployments, ensuring they are both robust and scalable.

Next, we will explore techniques specifically designed to enhance the overall environment of your Kubernetes systems. This will include optimizing cluster performance, embracing cutting-edge monitoring solutions, and exploring the integration of Kubernetes within diverse computing contexts such as edge environments and the **Internet of Things** (**IoT**). By building on the best practices, these upcoming discussions are geared toward fostering a culture of continuous improvement and innovation in your Kubernetes strategies.

# Enhancing the Kubernetes environment

Enhancing the overall stability and efficiency of Kubernetes operations is a critical aspect of modern cloud-native infrastructure management, especially while dealing with anti-patterns. This initiative delves into a range of strategic approaches and methodologies designed to bolster the robustness and operational efficacy of Kubernetes environments. It encompasses a holistic view of system optimization, covering everything from performance tuning to advanced resource management.

## Environment health checks and diagnostics

Conducting health checks and diagnostics in Kubernetes is a technical process involving specific tools and methodologies designed to ensure the cluster operates efficiently and reliably. This process is fundamental for early detection and resolution of issues, contributing significantly to the overall health of the Kubernetes environment.

## Health checks in Kubernetes

Kubernetes ensures the proper functioning and availability of applications through various health check mechanisms. These checks help to monitor and maintain the health of the components within a cluster. The following are key instances of how Kubernetes manages these checks:

- **Liveness and readiness probes**: Kubernetes uses liveness and readiness probes to check the health of Pods. Liveness probes determine if a Pod is running and functional. If a liveness probe fails, Kubernetes restarts the container. Readiness probes assess if a Pod is ready to receive traffic, ensuring that services don't route traffic to Pods that aren't ready.

- **Container health checks**: Containers within Pods can be configured with health checks using commands or HTTP requests. These checks are periodically performed to ensure the container is operational. If a container fails its health check, it can be automatically restarted by Kubernetes.

- **Node health status**: Kubernetes regularly checks the health of nodes in the cluster. The Node Controller in the Kubernetes Control Plane is responsible for monitoring the status of nodes. If a node becomes unresponsive, the Node Controller will mark it as unreachable, and the Scheduler will start rescheduling the affected Pods to other nodes.

- **Diagnostics in Kubernetes**: Conducting diagnostics in Kubernetes is a multifaceted technical process that involves monitoring, logging, event tracking, and direct interaction with the cluster's components. These activities are integral to identifying and resolving issues, ensuring the cluster remains healthy and performs optimally.

- **Logging and log analysis**: Kubernetes does not provide a native log storage solution, but it enables log aggregation at the cluster level. Tools such as Fluentd can be used to collect logs from various components and Pods. These logs can then be analyzed using solutions such as Elasticsearch and Kibana to identify issues and trends.

- **Monitoring tools**: Tools such as Prometheus are used to collect and record real-time metrics from the Kubernetes Control Plane and workloads running on the cluster. This data is crucial for diagnostics and can be visualized using platforms such as Grafana.

- **Event tracking**: Kubernetes generates events for significant state changes and errors. Tracking these events helps in diagnosing issues. Events can be viewed through the Kubernetes Dashboard or via the `kubectl` command-line tool.

- **Tracing and profiling**: For in-depth diagnostics, especially in microservices architectures, distributed tracing tools such as Jaeger or Zipkin can be used. These tools help trace the flow of requests through the microservices and identify bottlenecks or failures.

- **Debugging pods and services**: Kubernetes offers several commands such as `kubectl logs` for fetching logs of a container, `kubectl describe` to get detailed information about Kubernetes objects, and `kubectl exec` to execute commands in a container. These tools are essential for real-time diagnostics.

- **Network diagnostics**: Tools such as Cilium or Calico, which offer network observability features, can be used to diagnose networking issues within the cluster. They provide visibility into network policies, traffic flow, and potential network-related issues.

- **Performance monitoring**: Continuously monitoring the performance of applications and resources in Kubernetes is crucial. This involves tracking metrics such as CPU and memory usage, disk I/O, and network bandwidth.

## Stability enhancements

Stability enhancements in Kubernetes are crucial for ensuring that the system remains resilient and reliable under various operational conditions. These enhancements involve a series of technical strategies and configurations designed to fortify the Kubernetes environment against potential failures, disruptions, and performance issues. The aim is to create a Kubernetes setup that not only performs efficiently but also maintains its stability, even in the face of unexpected challenges.

### Pod and application stability

Kubernetes offers several mechanisms to promote the stability and reliability of applications running within pods. By utilizing these tools, Kubernetes can ensure that applications remain available and performant, even under varying loads and potential failures. Here's how Kubernetes achieves this:

- **ReplicaSets and Deployments**: Using ReplicaSets and Deployments is key to maintaining application stability. These ensure that a specified number of pod replicas are always running. If a pod fails, the ReplicaSet automatically creates a new one to replace it.

- **Liveness and readiness probes**: Configuring liveness and readiness probes helps Kubernetes determine the health and operational state of applications running in pods. These probes ensure traffic is only sent to healthy pods and restart those that have become unresponsive.

### Cluster-level stability

Kubernetes provides comprehensive tools and mechanisms to enhance the stability of the entire cluster. By actively managing the infrastructure and resources, Kubernetes helps ensure that the system remains resilient and efficient, ready to adapt to various operational demands and conditions. Here's how it achieves this:

- **Node health monitoring**: Regularly monitoring the health of nodes is essential. Kubernetes performs node health checks to detect and handle failed nodes. Pods running on an unhealthy node are automatically rescheduled to healthy ones.

- **Autoscaling**: Implementing HPA and Cluster Autoscaler ensures the cluster scales resources appropriately based on demand, contributing to overall stability by preventing resource exhaustion.

## *Networking and communication stability*

Maintaining robust and secure network operations is crucial for the uninterrupted functioning of services within a Kubernetes cluster. By setting stringent network policies and utilizing sophisticated service meshes, Kubernetes ensures that communication between services is seamless and stable:

- **Robust network policies**: Implementing comprehensive network policies in Kubernetes helps control the flow of traffic and can prevent network overloads or disruptions.

- **Service mesh implementation**: Utilizing service meshes such as Istio or Linkerd can greatly enhance stability. They provide advanced traffic management capabilities, including retries, circuit breaking, and sophisticated load balancing, which are vital for stable inter-service communication.

## *Operational and procedural stability*

Ensuring the operational and procedural integrity of a Kubernetes cluster is vital for sustained stability and security. Regular updates, comprehensive DRPs, and proactive management are essential components of a robust operational strategy:

- **Regular updates and patching**: Keeping the Kubernetes cluster and its applications updated with the latest patches is critical for stability. Regular updates ensure that the cluster is protected against known vulnerabilities and bugs.

- **DR planning**: Having a solid DRP, including regular backups and clearly defined recovery procedures, ensures the cluster can be quickly restored to a stable state after any disruptive event.

Kubernetes environments can be enhanced to achieve greater stability. This involves not just technical configurations and tools but also adhering to best practices in operational management. The goal is to build a Kubernetes ecosystem that can withstand fluctuations in workloads, infrastructure changes, and potential failures, thereby ensuring uninterrupted and stable operations.

# Enhancing data management and storage

Enhancing data management and storage in Kubernetes is a critical aspect of ensuring that applications run efficiently and reliably. As Kubernetes environments grow more complex, particularly with stateful applications, the need for sophisticated data management and robust storage solutions becomes paramount. This enhancement is focused on optimizing data storage, ensuring data persistence, and maintaining data integrity across the Kubernetes ecosystem.

## Persistent storage and dynamic provisioning

Kubernetes supports complex storage needs by providing robust solutions for persistent storage and dynamic provisioning. These features allow applications to efficiently manage storage resources, ensuring data persistence across pod restarts and deployments:

- **PVs and Persistent Volume Claims (PVCs)**: Utilizing PVs and PVCs effectively is key to managing storage in Kubernetes. PVs provide a way to allocate storage resources in a cluster, while PVCs allow applications to claim this storage. This setup separates storage configuration from its use, providing flexibility and ease of management.

- **Dynamic volume provisioning**: Implementing dynamic provisioning allows Kubernetes to automatically create storage resources as needed. This is achieved through StorageClasses, which define different types of storage offered in the cluster.

## Storage performance optimization

Optimizing storage performance is crucial for applications that demand high throughput and low latency. Kubernetes offers various options and configurations to fine-tune storage performance according to the specific needs of applications:

- **Choosing the right storage backend**: Depending on the application's needs, select the appropriate storage backend. Options include block storage for databases or file storage for shared filesystems. Cloud-native environments often leverage cloud provider-specific storage solutions for better integration and performance.

- **Fine-tuning storage parameters**: Optimize storage performance by fine-tuning parameters such as **input/output operations per second** (**IOPS**) and throughput. This involves understanding the application's storage access patterns and configuring the storage system accordingly.

## Data redundancy and replication

To ensure data availability and reliability, Kubernetes supports various data redundancy and replication strategies. These strategies help protect data against hardware failures and ensure it is available when needed:

- **HA configurations**: Ensure the HA of data by implementing replication strategies. This can be done within the storage layer, such as using **Redundant Array of Independent Disks** (**RAID**) configurations, or at the application level, such as database replication across multiple Pods.

- **Cross-region data replication**: In cloud environments, consider replicating data across multiple regions for DR and data locality.

## Backup and restore mechanisms

Regular backups and efficient restore processes are fundamental to safeguarding data in Kubernetes environments. Kubernetes supports various tools and strategies for backing up and restoring data, ensuring **business continuity (BC)**:

- **Regular data backups**: Implement regular backup processes for critical data. Tools such as Velero can be used for backing up Kubernetes resources and PVs.

- **Efficient restore processes**: Ensure that the backup solutions support efficient and reliable restore processes. Regularly test these processes to guarantee data can be restored quickly and accurately when needed.

## Data security and compliance

Maintaining data security and compliance is a top priority in Kubernetes deployments. Kubernetes offers features to help encrypt data and manage access, ensuring that sensitive information is protected against unauthorized access:

- **Encryption**: Encrypt sensitive data at rest and in transit. Kubernetes supports encrypting data at rest in `etcd`, and many storage backends offer built-in encryption capabilities.

- **Access controls**: Implement proper access controls to storage resources using Kubernetes RBAC and network policies to restrict access to sensitive data.

## Monitoring and management

Effective monitoring and lifecycle management of storage is essential for maintaining optimal performance and cost-efficiency in Kubernetes environments. Kubernetes provides tools to monitor storage utilization and manage the lifecycle of data:

- **Storage resource monitoring**: Monitor storage usage and performance metrics to proactively address capacity issues and performance bottlenecks

- **Lifecycle management**: Implement policies for data retention, archival, and deletion, particularly for meeting compliance requirements and managing costs in cloud environments

We discussed the various strategies employed to optimize Kubernetes, including cluster optimization, advanced monitoring, and integration with cloud and IoT. We also covered the importance of addressing security and multi-tenancy challenges, and the potential of leveraging AI and ML for continuous improvement and effective scaling.

# Summary

This chapter, *Practical Solutions and Best Practices*, provided an in-depth exploration of strategies for optimizing Kubernetes environments while addressing common anti-patterns. It offered a combination of technical solutions and operational best practices aimed at enhancing the efficiency, stability, and resilience of Kubernetes deployments.

The chapter emphasized a holistic approach to management by integrating technical skills with strategic planning. It highlighted the importance of continuous monitoring and adaptation to Kubernetes' evolving ecosystem. Additionally, it focused on efficient management and the necessity of a deep understanding of Kubernetes to fully harness its capabilities.

In the next chapter, the focus shifts toward implementing insights and solutions derived from these case studies across various sectors. It explores advanced strategies for ensuring sustainable IT practices and discusses the long-term impacts of these improvements.

# 5
# Real-World Case Studies

This chapter presents a series of real-world case studies that illustrate the challenges and solutions associated with Kubernetes anti-patterns. Through the lens of actual organizational experiences, it highlights the journey that must be undertaken, from encountering operational pitfalls to implementing strategic solutions. The narratives cover a spectrum of industries and issues, from a tech startup's resource over-provisioning to security enhancements in banking, offering insights into the practical application of Kubernetes best practices. Each study underscores the importance of tailored strategies in overcoming specific obstacles, paving the way for future advancements, and setting a precedent for operational excellence in Kubernetes environments.

We will cover the following topics in this chapter:

- Learning from real organizations' experiences
- Case studies on anti-patterns and solutions
- Future directions

## Learning from real organizations' experiences

My experience as a Kubernetes consultant has allowed me to witness firsthand the transformative effects of addressing these anti-patterns. I recount the tales of businesses that recognized the pitfalls of their initial Kubernetes strategies – stories infused with the challenges of adapting to a system that promises as much complexity as it does utility.

I recall the early days of engaging with a fledgling tech startup. They were enthusiastic yet ensnared in the common trap of over-provisioning. Guiding them through a strategic scaling back, we discovered the delicate balance between resource availability and cost-effectiveness. It was a formative lesson in the nuanced art of resource management within Kubernetes.

Then, there was the major retail corporation, buckling under the weight of traffic during peak seasons. Collaboratively, we unraveled their load balancing woes, crafting a solution that not only stabilized their online platform but also enhanced customer satisfaction. This experience sharpened my understanding of the critical role of responsive load management in Kubernetes environments.

My involvement with the healthcare sector brought to light the paramount importance of data integrity and compliance within Kubernetes-managed storage systems. Working closely with them to revamp their persistent storage strategies, I learned the intricacies of aligning technical infrastructure with rigorous regulatory demands.

Each organization's story has been a chapter in my professional growth, contributing to a reservoir of knowledge that I draw upon to this day. From enhancing security measures within the banking industry to streamlining deployment processes in manufacturing, every challenge that I've surmounted has been a stepping stone to greater expertise.

As we navigate each case, we'll see patterns emerge – common threads that tie these varied experiences together. These are the lessons that forge stronger architects, developers, and administrators, equipping them with the foresight to anticipate and nullify anti-patterns before they take root.

In sharing these experiences, I aim not only to impart lessons learned but also to demonstrate the growth potential that lies in each Kubernetes deployment. Whether it's reducing microservice dependencies in telecommunications or improving autoscaling in educational institutions, these real-world experiences have honed my skills and shaped my approach as a Kubernetes expert. They are a reminder that beyond the technical solutions, it's the journey of learning and adaptation that truly transforms organizations.

# Case studies on anti-patterns and solutions

In this section, we'll cover a few use cases to understand the problem and possible solutions and lessons we can learn from it.

## Use case 1 – a fintech startup overcomes over-provisioning resources through strategic solutions

**Background**:

A burgeoning startup in the fintech sector sought to carve out its niche by offering cutting-edge payment processing services. In its quest to ensure high availability and fault tolerance, the startup aggressively over-provisioned resources within its Kubernetes clusters. This approach led to a significant surge in operational costs, which began to erode the company's capital reserves and impede its ability to invest in other critical areas, such as research and development and customer acquisition.

**Problem statement**:

As the user base grew, the workload demands became more unpredictable, and the startup found that its static resource allocation strategy was neither sustainable nor cost-effective. The Kubernetes cluster was often idle during off-peak hours, but the resources were still reserved and accruing expenses. Moreover, during unexpected spikes in demand, the manual scaling processes were too slow, resulting in performance bottlenecks that affected end user experience.

The realization dawned upon the startup's leadership that their Kubernetes infrastructure, while robust, was not optimized. It was evident that to maintain its competitive edge and financial health, the startup needed to address the anti-pattern of over-provisioning resources. The challenge was to implement a resource allocation strategy that could dynamically adapt to fluctuating workloads, optimize costs, and maintain the highest service levels required by financial service standards.

The problem was multifaceted:

- **Cost inefficiency**: The financial overhead of maintaining surplus capacity was unsustainable, especially for a startup operating within the capital-intensive fintech industry

- **Resource underutilization**: A significant portion of computational resources was underutilized, leading to wasted expenditure without corresponding business value

- **Scalability lag**: The inability to scale resources promptly in response to varying loads compromised performance during critical periods

- **Complexity in management**: Manual intervention for scaling and resource allocation was prone to human error and was not viable long-term as the company aimed to scale its operations

**Solution implementation**:

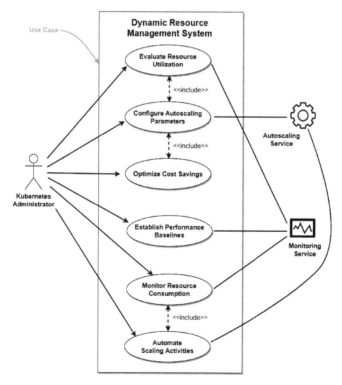

Figure 5.1 – Dynamic resource management system solution

The solution depicted in the preceding use case diagram revolves around a dynamic resource management system that addresses the resource allocation inefficiencies in a fintech startup's Kubernetes clusters. The Kubernetes administrator initiates this process by evaluating the current resource utilization across the system. This evaluation is crucial to understanding where resources are being used effectively and where they aren't.

Autoscaling parameters are then configured to align resource provisioning with the actual workload demands. These parameters enable the system to automatically scale resources up during high-traffic periods, ensuring that customer transactions are processed efficiently. Conversely, the system scales down during periods of low activity to prevent unnecessary expenditure on idle resources. This scaling is managed by the autoscaling service, which adjusts resources in real time based on the workload.

The monitoring service supports these operations by providing ongoing oversight of resource consumption. It ensures that the autoscaling service has the most current information on system demands, enabling precise scaling actions.

**Results**:

Together, these components work in tandem to create a responsive and cost-efficient infrastructure, dynamically adapting to the fluctuating needs of the startup's operations without the need for constant manual adjustments. This system not only minimizes the risk of performance issues during critical periods but also optimizes the startup.

## Use case 2 – improving load balancing in a major retail corporation

**Background**:

The retail industry thrives on its ability to provide seamless customer service, particularly during peak shopping seasons. A major retail corporation, with a significant online presence and a vast array of products, faced critical challenges with its load balancing mechanisms within its Kubernetes infrastructure. The corporation's online platform experienced heavy and unpredictable traffic, which was exacerbated during sales events and holidays.

**Problem statement**:

Their existing load balancing solution was static and unable to efficiently distribute traffic among the available nodes, leading to server overloads and subsequent downtime.

This inefficient load balancing resulted in several detrimental effects:

- **Customer service disruption**: During traffic spikes, customers experienced slow response times and, in worst cases, service outages, directly impacting customer satisfaction and trust

- **Sales losses**: Every minute of downtime translated into substantial financial loss due to interrupted transactions and abandoned shopping carts

- **Strained infrastructure**: Certain nodes were consistently overburdened, while others remained underutilized, leading to uneven wear and potential early failure of hardware

- **Operational inefficiency**: The IT team spent considerable time firefighting issues related to traffic surges instead of focusing on strategic initiatives

Leadership recognized that the corporation's Kubernetes-based platform required a dynamic and intelligent load balancing solution that could not only respond to current demand but also predict and scale according to future traffic patterns. The challenge encompassed not just the implementation of a more responsive load balancing system, but also the integration of this system with their existing Kubernetes setup without disrupting ongoing operations.

**Solution implementation**:

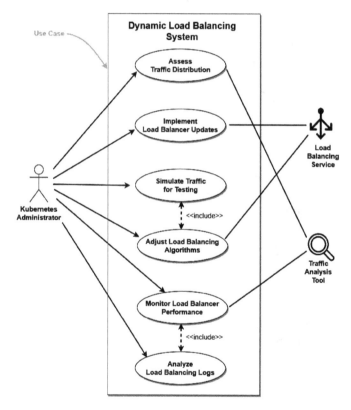

Figure 5.2 – Dynamic load balancing system

The Kubernetes administrator is central to this solution, spearheading the initiative to improve how the online platform handles incoming traffic. This individual starts by assessing traffic distribution to understand where bottlenecks are forming and which nodes are under or over-utilized.

Following this assessment, the Kubernetes administrator implements updates to the load balancer, likely involving the introduction of more dynamic and responsive load balancing algorithms that can adapt to traffic in real time. This task is crucial for preventing server overloads during unexpected surges in user activity.

To ensure these new algorithms work as intended, the administrator simulates traffic, creating a controlled testing environment to observe how the updated load balancer performs under various conditions. This step is vital for validating the effectiveness of the load balancing strategy before it goes live.

The load balancing service is an automated system that actively manages the distribution of traffic across the platform's nodes. It works hand-in-hand with the Kubernetes administrator's configurations to ensure that resources are allocated efficiently.

Monitoring performance is a continuous process, as reflected in the use case diagram. The performance of the load balancer is tracked to ensure that the newly implemented strategies effectively mitigate the previous issues of slow response times and outages.

Finally, the traffic analysis tool plays a supporting role by providing detailed insights into traffic patterns. This tool enables data to be collected that feeds into the continuous improvement of load balancing strategies.

**Results**:

By analyzing the load balancing logs, the system can learn from past performance, identifying successful configurations and areas for further optimization. This data-driven approach ensures that the system becomes progressively more attuned to the corporation's specific traffic patterns and demands.

## Use case 3 – resolving persistent storage issues in the healthcare sector

**Background**:

A Kubernetes-driven IT environment within the healthcare sector faced critical challenges with persistent storage – a fundamental requirement for maintaining electronic health records and supporting real-time patient care systems. The sector's reliance on Kubernetes was grounded in its need for high availability and scalability.

**Problem statement**:

The persistent storage solution in place was falling short of the sector's stringent data management and regulatory compliance requirements.

The persistent storage issues were manifesting in several ways:

- **Data integrity risks**: Inconsistent data replication and backup strategies were leading to concerns about data integrity and potential loss, which could have dire consequences for patient care

- **Access delays**: Slow retrieval times for medical records were impeding healthcare providers' ability to access vital patient information promptly

- **Scalability bottlenecks**: As the volume of data grew, the existing storage solution struggled to scale efficiently, leading to performance degradation

- **Compliance concerns**: The inability to guarantee data availability and integrity raised serious compliance issues with healthcare regulations

With a growing patient database and an ever-increasing reliance on digital solutions, resolving these persistent storage issues was not just a matter of operational efficiency but also of patient safety and regulatory compliance. The challenge was to overhaul the Kubernetes persistent storage strategy without disrupting the critical services that patients and healthcare providers depended on daily.

**Solution implementation**:

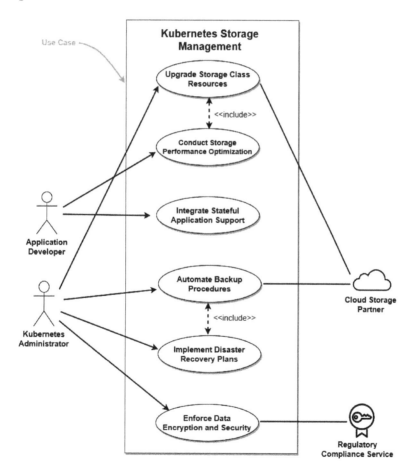

Figure 5.3 – Kubernetes persistent storage strategy

The preceding use case diagram illustrates a comprehensive approach to revamping the persistent storage strategy. The aim is to create a system that ensures high availability, scalability, and compliance with strict data management regulations necessary for patient care.

At the core of this strategy, the Kubernetes administrator is tasked with upgrading storage class resources to meet the growing data demands and ensure that the storage solution can scale effectively. This upgrade is a pivotal step in maintaining data integrity and ensuring that healthcare providers have quick access to medical records.

The administrator also works on optimizing storage performance, which is essential for handling the large volumes of sensitive data that the healthcare sector deals with daily. This optimization helps address the scalability bottlenecks that have previously led to performance issues.

Integrating support for stateful applications is another crucial element, ensuring that applications that require persistent storage can function reliably within the Kubernetes environment. This integration is vital for applications handling electronic health records and patient care systems, where data persistence is non-negotiable.

Automating backup procedures is implemented to protect against data loss. These automated processes are designed to ensure that data replication and backups are performed consistently, safeguarding data integrity.

Disaster recovery plans are put in place as a precautionary measure. These plans provide a clear protocol for restoring data and services in the event of a system failure, which is essential for maintaining continuous patient care.

**Results**:

Enforcing data encryption and security is an integral part of the strategy to comply with healthcare regulations and protect patient information. This step ensures that all data, at rest or in transit, is encrypted securely, addressing compliance concerns and safeguarding against unauthorized access.

The cloud storage partner and regulatory compliance service are external entities that provide support and oversight. The cloud storage partner offers scalable storage solutions and backup services, while the regulatory compliance service ensures that the storage strategy adheres to healthcare regulations.

# Use case 4 – enhancing cluster security in a small finance bank

**Background:**

Security is the cornerstone of the banking industry, which is increasingly reliant on technology to manage assets, transactions, and customer data. A notable trend within the industry has been the adoption of Kubernetes to orchestrate containerized applications. However, this transition has not been without its challenges. One of the most pressing issues was the need to enhance cluster security to safeguard against both external breaches and internal vulnerabilities.

**Problem statement:**

The bank's Kubernetes clusters were facing several security concerns:

- **Vulnerability to cyber threats**: With the increasing sophistication of cyberattacks, the existing security measures within the clusters were proving inadequate, risking financial data and customer trust

- **Compliance and regulatory hurdles**: Banks are subject to stringent regulatory requirements, and the existing Kubernetes configuration wasn't fully compliant, potentially leading to legal and financial repercussions

- **Insider threats and misconfigurations**: There was an urgent need to mitigate risks arising from internal misconfigurations and insider threats, which could lead to unauthorized access or data leaks

- **Incident response and forensics**: The existing infrastructure lacked robust mechanisms for incident response and forensic analysis, which is critical for addressing breaches and understanding attack vectors

The stakes were incredibly high; any security lapse could result in significant financial loss, erosion of customer confidence, and severe regulatory penalties. The challenge for the bank was to implement a cluster security framework that was comprehensive, agile, and fully integrated with Kubernetes' dynamic nature, all while maintaining uninterrupted financial services.

**Solution implementation:**

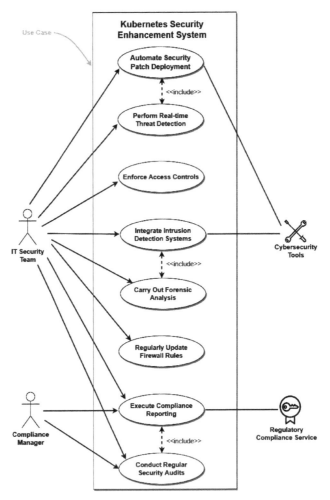

Figure 5.4 – Kubernetes security system enhancement

The preceding use case diagram illustrates a strategic approach to enhancing the security framework of Kubernetes clusters. It represents an action plan to safeguard against cyber threats, ensure compliance with stringent regulatory standards, and establish robust incident response protocols.

The IT security team begins by automating the deployment of security patches, ensuring that the system is promptly and consistently protected against known vulnerabilities. Real-time threat detection is also implemented, providing the team with immediate alerts to potential security breaches, thus allowing for swift action.

Access controls are rigorously enforced to maintain a secure environment, restricting unauthorized access and mitigating insider threats. This is complemented by the integration of intrusion detection systems, which monitor the network for signs of compromise, feeding into the proactive security posture of the bank.

Forensic analysis capabilities are developed to delve into security incidents, uncovering the root causes and preventing recurrence. This forensic readiness ensures that the bank can quickly recover from an incident and provides evidence for any required legal proceedings.

The compliance manager oversees the execution of compliance reporting, a critical aspect that ensures the bank meets all regulatory obligations. Regular security audits are conducted to review the effectiveness of security measures and compliance adherence.

**Results**:

Supporting these activities are external cybersecurity tools that provide advanced capabilities for threat detection, analysis, and response. The regulatory compliance service plays an advisory role, ensuring that all security measures align with the latest regulations and industry best practices.

## Use case 5 – addressing inadequate monitoring in an e-commerce giant

**Background**:

For an e-commerce giant, maintaining system reliability and customer satisfaction is paramount, and this hinges on the ability to monitor complex distributed systems effectively. Unfortunately, this enterprise found itself ensnared in several monitoring anti-patterns within its Kubernetes environment. Reliance on legacy monitoring tools, inadequate alert configurations, and a lack of actionable insights from gathered data led to a reactive rather than proactive approach to system health and performance.

**Problem statement**:

The following key anti-patterns were plaguing the e-commerce giant's Kubernetes setup:

- **Silent failures**: Critical failures were slipping through undetected, only coming to light through customer complaints rather than internal alerts

- **Alert fatigue**: The inundation of non-critical alerts desensitized the operations team to warnings, allowing significant issues to go unrecognized amidst the noise

- **Manual correlation**: The lack of intelligent automation forced teams to manually correlate data across systems to diagnose issues, leading to delays and potential human error

- **Performance blind spots**: Key performance indicators were not adequately monitored, creating blind spots in understanding the customer experience and system efficiency

The e-commerce giant faced the dual challenge of overhauling its monitoring infrastructure to escape these anti-patterns and doing so in a manner that scaled across its global operations without disrupting ongoing services.

**Solution implementation:**

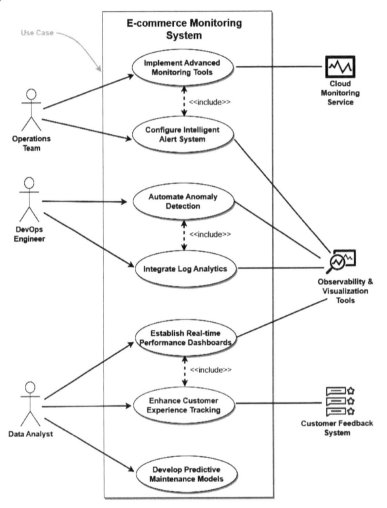

Figure 5.5 – E-commerce monitoring system

The preceding use case diagram illustrates an upgraded monitoring system for an e-commerce giant, which is tackling the intricate challenge of effectively monitoring its distributed systems within a Kubernetes environment. The strategy focuses on transitioning from a reactive to a proactive monitoring posture, addressing the silent failures, alert fatigue, manual correlations, and blind spots that have been impacting system reliability and customer satisfaction.

The operations team is at the forefront, integrating advanced monitoring tools that provide deeper visibility into the system's operations. This integration allows for a more nuanced detection of issues, ideally preventing problems before they affect customers.

To combat the deluge of non-critical alerts that have led to alert fatigue, the team sets up an intelligent alert system designed to prioritize alerts. This ensures that the most critical issues are flagged for immediate action, reducing the noise and helping the team to focus on genuinely impactful system events.

The DevOps engineer takes charge of implementing automation for anomaly detection, which is crucial for quickly identifying and responding to unexpected system behavior without the need for labor-intensive manual data analysis.

With the integration of comprehensive log analytics, the system gains the capability to perform in-depth analysis and correlation of logs across different services, which is key in diagnosing complex issues that may span multiple components of the infrastructure. This integration is crucial in transitioning away from the previously manual and error-prone correlation process.

Data analysts bring their expertise to bear by establishing real-time performance dashboards, providing a live view of the system's health and efficiency. These dashboards are critical in illuminating performance metrics that were previously not monitored adequately, helping to identify and resolve any issues affecting customer experience.

To further hone in on customer satisfaction, measures to enhance customer experience tracking are put into place. This enables the e-commerce company to capture and analyze customer feedback and behavior, ensuring that the digital experience aligns with customer expectations and needs.

Predictive maintenance models are also developed by the team. These models leverage historical data to forecast potential system issues, allowing for preventative maintenance and reducing the likelihood of unexpected downtime.

**Results**:

Supporting the internal team's efforts, external services such as the cloud monitoring service and observability and visualization tools provide additional layers of monitoring and data visualization capabilities. These services supplement the company's monitoring efforts, offering scalability and advanced analytical tools. Furthermore, a customer feedback system is integrated to gather direct input from users, which can inform continuous improvements in terms of system performance and user experience.

## Use case 6 – streamlining complex deployments in a manufacturing company

**Background**:

A manufacturing company utilizing Kubernetes for orchestrating its applications faced a common anti-pattern of the complication of deployment workflows. With a multi-faceted infrastructure to support various stages of production, the Kubernetes deployment processes became increasingly convoluted. This complexity not only slowed down the deployment of new applications and updates but also increased the risk of errors, which could lead to production halts or defects in the manufacturing pipeline.

**Problem statement**:

The complexity of the Kubernetes deployment workflows manifested in several problematic ways:

- **Deployment bottlenecks**: Overly complex deployment processes created bottlenecks, causing significant delays in rolling out new features and updates

- **Increased risk of downtime**: Each deployment carried a high risk of errors, with the potential to disrupt manufacturing operations, leading to costly downtime

- **Resource mismanagement**: The inefficient deployment patterns led to poor utilization of computational resources, resulting in unnecessary overheads

- **Operational overhead**: The IT team's operational load increased as they navigated the cumbersome deployment process, diverting attention from innovation and optimization efforts

Faced with the need to streamline its Kubernetes deployment processes, the manufacturing company embarked on a strategic initiative to re-engineer its deployment pipelines. The goal was to adopt a more straightforward, automated, and error-proof deployment strategy that aligns with the just-in-time principles of modern manufacturing.

**Solution implementation:**

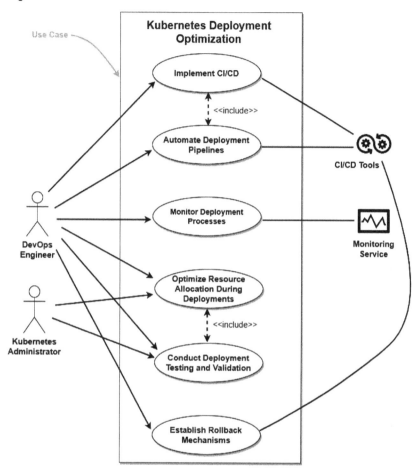

Figure 5.6 – Kubernetes deployment automation

The plan starts with the DevOps engineer, who implements **continuous integration/continuous deployment (CI/CD)**, which is a method to automate the deployment pipelines. This automation ensures that new applications and updates are delivered more efficiently, helping to prevent the slowdowns that were previously occurring.

To support this, the automated deployment pipelines are crucial as they enable consistent and error-free deployments, directly addressing the potential for production disruptions.

The monitoring service is an integral part of the strategy, providing visibility into each deployment process. This visibility is key to preventing downtime as it allows for immediate detection and resolution of any issues that arise during deployment.

The Kubernetes administrator focuses on optimizing resource allocation during deployments, which is essential for the efficient use of computational resources and the avoidance of unnecessary expenses.

To ensure that each deployment meets quality standards, the team carries out thorough testing and validation. This step is fundamental in catching any issues before they can affect the production environment.

**Results**:

This establishes the safety features that allow the system to revert to a stable state if a deployment introduces errors, ensuring the continuity and stability of manufacturing operations.

## Use case 7 – managing resource limits in a national media company

**Background**:

A national media company, with a vast digital presence and a significant volume of daily content updates, faced a critical Kubernetes anti-pattern of improper management of resource limits. This mismanagement led to several issues within their Kubernetes environment, from inefficient resource utilization to critical application failures during peak news cycles. Without clearly defined resource requests and limits, the Kubernetes scheduler was unable to effectively allocate resources across the company's pods and nodes, resulting in both resource starvation and overcommitment.

**Problem statement**:

The consequences of not managing Kubernetes resource limits effectively were multifaceted:

- **Service instability**: Inadequately set resource limits caused pods to either be killed for exceeding limits or underperform due to insufficient resources, leading to service disruptions
- **Inconsistent application performance**: The lack of proper resource allocation resulted in unpredictable application performance, with some services running sluggishly while others hoarded unused resources
- **Cost inefficiencies**: The company was incurring unnecessary costs by overprovisioning resources to avoid service disruptions, leading to significant financial waste
- **Compromised scalability**: The ability to scale services dynamically in response to viewership demand was hindered, affecting the company's agility and responsiveness to breaking news events

The national media company's challenge was to implement a resource management strategy that could dynamically adjust to the load imposed by breaking news and fluctuating viewership while optimizing costs and maintaining high service availability.

**Solution implementation**:

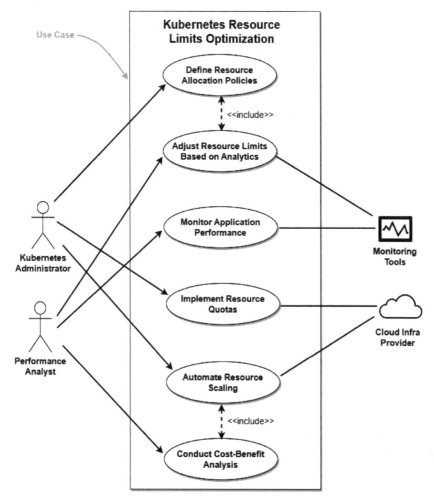

Figure 5.7 – Kubernetes resource limits optimization

The Kubernetes administrator is tasked with defining clear resource allocation policies. These policies will guide how resources are distributed among the company's applications, ensuring that each component has access to the resources it needs without wasting any.

Based on the insights provided by performance analytics, the administrator can adjust resource limits to match the actual usage patterns. This flexibility is critical during peak news cycles, where viewership can fluctuate dramatically, and resources need to be allocated or de-allocated quickly.

Monitoring application performance is an ongoing process that's aided by sophisticated monitoring tools. These tools provide real-time insights into how applications are performing and how resources are being used, enabling proactive management of resource allocation.

Implementing resource quotas is another step taken by the administrator. Quotas prevent any single application or service from using more resources than necessary, which helps to avoid over-commitment and ensures that resources are available for other services that might need them.

Automating resource scaling is a significant part of the strategy. This automation allows the system to respond swiftly to changes in demand, scaling up during high viewership and scaling down when demand drops, ensuring efficient use of resources and helping to manage costs.

A performance analyst conducts a cost-benefit analysis to evaluate the financial impact of resource allocation strategies.

**Results**:

This analysis helps to avoid financial waste by ensuring that resource use is aligned with the company's budget and value derived from resource expenditure.

External services, such as cloud infrastructure providers, offer scalable resource options and can be leveraged to extend the company's capacity quickly when needed.

## Use case 8 – reducing microservice dependencies in telecommunications

**Background**:

In the fast-paced world of telecommunications, the ability to rapidly adapt and scale services is crucial. A prominent telecommunications company, leveraging Kubernetes to manage its microservices architecture, encountered a significant anti-pattern: excessive interdependencies among its microservices.

**Problem statement**:

This tangled web of dependencies led to a complex and fragile system architecture, where changes in one service could inadvertently impact others, causing stability issues and hindering the deployment of new features.

Here are some of the challenges that were stemming from these microservice dependencies:

- **Deployment complexity**: The interdependent nature of services made deployments cumbersome and risky as a single change could potentially disrupt multiple services
- **Difficulty in isolating failures**: When issues occurred, it was challenging to pinpoint and isolate them due to the intricate dependency chains, leading to prolonged downtimes

- **Scalability hurdles**: Scaling individual services became problematic as it required careful coordination to ensure that dependent services were not adversely affected

- **Inhibited innovation**: The fear of causing widespread issues led to a reluctance to update or improve individual services, thereby stifling innovation and progress

Confronted with the need to simplify and decouple its microservices, the telecommunications company decided on a strategic shift of its Kubernetes environment. The objective was to restructure the microservices architecture to reduce dependencies, thereby enhancing system stability, scalability, and agility.

**Solution implementation:**

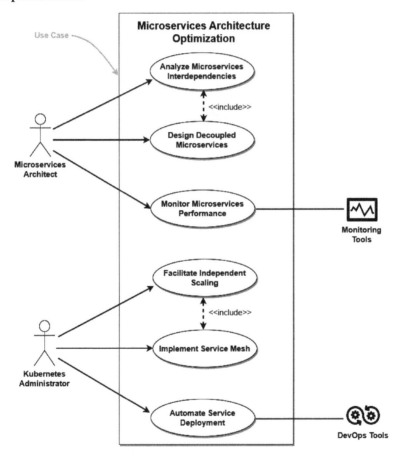

Figure 5.8 – Microservices architecture optimization

The microservices architect begins the optimization process by analyzing the existing interdependencies among microservices. This analysis is essential to understand the complex web of interactions and to identify which services are excessively reliant on one another.

Following this analysis, the architect designs decoupled microservices. By separating these services and reducing their interdependencies, the system's overall architecture becomes more robust and less prone to cascading failures that can occur when one service impacts another.

The Kubernetes administrator plays a crucial role in this strategy. They facilitate independent scaling of microservices, allowing each service to be scaled up or down based on its demand without affecting others. This independence is key to addressing the scalability hurdles previously faced.

The administrator also implements a service mesh, which is an infrastructure layer that allows for secure and efficient communication between different microservices. The service mesh aids in managing service interactions, providing more granular control and observability.

**Results**:

To streamline the deployment process, service deployment is automated with the help of DevOps tools. Automation ensures that deployments are consistent, repeatable, and less prone to human error, thereby reducing deployment complexity and the risk associated with manual deployments.

The performance of microservices is continuously monitored using sophisticated monitoring tools. These tools provide insights into how each microservice performs, allowing for the quick identification and isolation of any failures.

## Use case 9 – improving inefficient autoscaling in an educational institution

**Background**:

An educational institution utilizing Kubernetes faced a significant challenge with its existing autoscaling setup. The autoscaling mechanisms in place were inefficient, often leading to delayed scaling during critical periods such as online enrollment or e-learning sessions.

**Problem statement**:

This inefficiency not only impacted the user experience but also led to resource wastage during off-peak times.

The primary issues with the institution's Kubernetes autoscaling were as follows:

- **Delayed response to traffic surges**: The autoscaling system was slow to respond to sudden increases in demand, causing performance bottlenecks during peak usage times

- **Over-provisioning during low traffic**: Conversely, the system was slow to scale down resources when demand waned, leading to unnecessary resource utilization and associated costs

- **Lack of customized scaling metrics**: The autoscaling was primarily based on basic metrics such as CPU and memory usage, which didn't accurately reflect the needs of different applications run by the institution

- **Operational challenges**: The IT team faced difficulties in managing the scaling processes, which required frequent manual interventions and adjustments

The educational institution recognized the need to refine its autoscaling strategies to ensure that its digital learning platforms could handle variable loads reliably while optimizing resource usage.

**Solution implementation:**

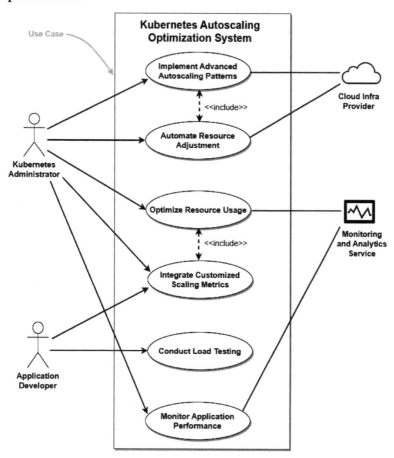

Figure 5.9 – Kubernetes autoscaling optimization system

The Kubernetes administrator is set to implement advanced autoscaling patterns. These patterns are more sophisticated than basic CPU and memory metrics and are designed to respond swiftly to changes in demand. This responsiveness is crucial during times such as online enrollment or e-learning sessions when the system must handle surges in user activity without delay.

Automation of resource adjustment is a key element of this strategy. By automating, the system can promptly scale up resources when there's a spike in demand and scale down when the demand drops, optimizing resource usage and preventing over-provisioning during periods of low traffic.

The administrator also integrates customized scaling metrics tailored to the specific needs of the educational institution's applications. Unlike the basic metrics that were used previously, these customized metrics provide a more accurate reflection of each application's resource requirements.

An application developer is involved in conducting load testing. This testing is essential to ensure that the autoscaling system performs as expected under various load conditions. Load testing helps to simulate both peak and off-peak scenarios, verifying that the autoscaling responds correctly.

**Results**:

The monitoring and analytics service continuously tracks the performance of applications, providing insights that inform further optimization of the autoscaling system.

# Use case 10 – correcting configuration drift in a major energy company

**Background**:

A leading energy company, utilizing Kubernetes to manage its diverse and expansive digital infrastructure, faced the common issue of configuration drift. This phenomenon, where configurations diverge or become inconsistent over time, was particularly problematic given the scale and complexity of the company's operations.

**Problem statement**:

This drift not only jeopardized system stability and performance but also posed significant risks in terms of regulatory compliance and security, both critical concerns in the energy sector.

Some of the challenges that were posed by configuration drift in the company's Kubernetes environment are as follows:

- **Deployment inconsistencies**: Disparities in environment configurations led to unpredictable behavior of applications across different stages, from development to production

- **Exposure to security threats**: Inconsistent application of security updates and patches across clusters heightened the risk of vulnerabilities and potential breaches

- **Compliance deviations**: The company, operating under strict regulatory standards, faced serious compliance risks due to these configuration inconsistencies

- **Resource-intensive rectification**: The effort that was required to identify, troubleshoot, and rectify configuration discrepancies consumed significant resources, impacting operational efficiency

Faced with these challenges, the energy company started systematically addressing the issue of configuration drift within its Kubernetes environment. The goal was to establish a mechanism that ensured consistency, security, and compliance across all deployments.

**Solution implementation:**

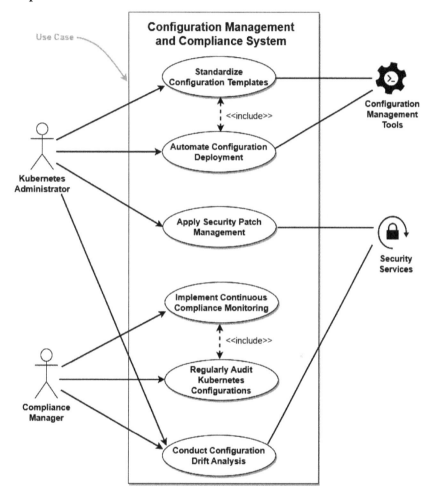

Figure 5.10 – Configuration management and compliance system

The Kubernetes administrator begins by standardizing configuration templates. These templates serve as blueprints for deployments, ensuring uniformity across the company's digital infrastructure. This standardization is key to reducing deployment inconsistencies and ensuring that applications behave predictably from development through to production.

To streamline the process, configuration deployment is automated, which helps maintain consistency as the infrastructure evolves. Automation ensures that security updates and patches are uniformly applied across all clusters, mitigating the risk of vulnerabilities that could lead to security breaches.

The compliance manager implements continuous compliance monitoring to ensure adherence to the stringent regulatory standards governing the energy sector. This ongoing monitoring is critical to identifying and addressing compliance deviations promptly.

Regular audits of Kubernetes configurations are also scheduled. These audits are essential in detecting configuration drift and identifying discrepancies between the current state and the standardized templates.

Conducting configuration drift analysis is another crucial action. It involves detailed inspections to understand the root causes of drift and to inform the development of strategies to prevent future occurrences.

**Results**:

This effort resulted in tools that provide the necessary technology to manage configurations at scale, and security services that offer specialized expertise in maintaining the security posture of Kubernetes environments.

Having explored various real-world scenarios where organizations have successfully navigated Kubernetes anti-patterns, we've seen firsthand how strategic solutions can transform potential setbacks into operational success. From tech startups to major retail corporations, each case study has provided a unique glimpse into overcoming specific Kubernetes challenges through innovative approaches and tailored solutions.

As we move forward, we will shift our focus to the future directions after these challenges have been addressed. The next section will discuss how organizations can continue to evolve and adapt their Kubernetes environments to stay ahead of the curve. We'll explore emerging trends, potential new challenges, and the ongoing development of Kubernetes capabilities to ensure that your infrastructure not only meets current needs but is also prepared for future demands.

## Future directions

As we look ahead after overcoming the challenges in past cases, with Kubernetes now strong and stable, companies can look forward to exciting possibilities.

Kubernetes will soon be a key player in digital transformation. Businesses that have improved their operations can now use Kubernetes to be more innovative. It will be the foundation for DevSecOps, where security is a part of the whole process, not just an afterthought.

Using microservices has shown us that being modular and separate is not just about design; it's also a smart business move. Kubernetes will continue helping companies grow these services separately. This means quicker and more targeted updates that can adapt to the market faster.

Data will be a big deal. Kubernetes will help organize complex data work that powers analytics and machine learning. Companies that have fixed resource problems will use Kubernetes to make their data systems better for real-time insights.

On the technical side, more tools will be added to the Kubernetes community. There will be new plugins and tools to make it easier to manage clusters and have more control over updates. These will be user-friendly and make Kubernetes easier for everyone to use.

Lastly, Kubernetes will work closely with cloud services. This will create new ways to use both public and private clouds, providing more flexibility and strength. Kubernetes, which has shown how good it is in single companies, will now be important in cloud-focused operations.

This path shows that Kubernetes is moving from just managing infrastructure to being a big part of making a company better and more innovative in a cloud-first world.

## Summary

This chapter unfolded the complexities of Kubernetes anti-patterns through real-world case studies, providing a window into the practical challenges and innovative solutions that are deployed by various organizations. It illustrated the significance of customized strategies to address unique operational issues, from resource allocation to security vulnerabilities. It even underlined Kubernetes' adaptability to diverse operational needs when wielded with expertise. It showcased the value of precise issue identification and the application of best practices tailored to specific industry demands. By offering a panoramic view of these case studies, it reinforced the concept that Kubernetes is not just a tool but a versatile platform that, when mastered, can significantly enhance system operations and efficiency.

In the next chapter, we'll explore the diverse techniques for optimizing Kubernetes performance and cover cluster resource allocation, image management, and network tuning. After, we'll explore strategies for enhancing scalability through design principles such as statelessness and adopting microservices architectures. Lastly, we'll examine maximizing Kubernetes' potential by integrating with cloud-native ecosystems, leveraging continuous deployment, and optimizing multi-cloud strategies. The next chapter also touches on cost management, the use of AI, and best practices for security.

# 6

# Performance Optimization Techniques

This chapter explores practical ways to boost performance and efficiency in Kubernetes environments. It covers a range of topics, from improving how resources are used and managed to getting the most out of container systems. The discussion includes optimizing network and storage performance, which are key to running Kubernetes smoothly.

This chapter also looks at how to scale systems effectively, touching on the use of microservices, cloud-native technologies, and modern approaches such as GitOps. Each area is broken down into understandable strategies and practices, providing valuable insights for those looking to build stronger, more efficient Kubernetes setups. This guide is an essential tool for anyone aiming to improve their Kubernetes operations and achieve high-level performance and efficiency.

We'll cover the following topics in this chapter:

- Techniques to optimize Kubernetes performance
- Ensuring efficiency and scalability
- Maximizing the potential of Kubernetes

## Techniques to optimize Kubernetes performance

This section explores strategies for enhancing Kubernetes performance by focusing on optimizing resource allocation, container management, network and data performance, and system health. It emphasizes efficient logging, monitoring, and load balancing to improve overall cluster functionality and efficiency.

## Evaluating cluster resource allocation

Kubernetes cluster resource allocation evaluation involves a detailed analysis of how resources are distributed and used across the cluster. It's a process that ensures that applications receive the necessary resources to perform effectively, without overburdening the cluster or wasting resources.

Here's a simplified breakdown of the process for evaluating cluster resource allocation in Kubernetes:

1.  **Understanding resource needs**: Assess how resources are allocated to ensure applications perform well without excess.

    **What it is**: It involves evaluating how resources such as CPU and memory are used in your Kubernetes cluster.

    **Why it matters**: It ensures each application has enough resources to function well without wasting capacity or overloading the cluster.

2.  **Collecting data**: Gather resource usage data to identify inefficiencies and opportunities for optimization.

    **Tools used**: Kubernetes Metrics Server for basic metrics and Prometheus and Grafana for more detailed insights and visualizations.

    **Purpose**: To track how resources are currently being used, helping to spot any issues, such as resource contention (where apps fight over resources) or underutilization (where resources aren't fully used).

3.  **Analyzing resource allocation**: Review and adjust resource settings to optimize the balance between application needs and cluster resources.

    **Checking settings**: Review the resource requests and limits set for pods and containers:

    *   **Requests**: Ensure each application has the minimum resources needed to run

    *   **Limits**: Prevent any application from using more than its fair share, protecting others

    **Importance**: Proper settings help the Kubernetes scheduler efficiently place pods on nodes, balancing application needs with available cluster resources.

4.  **Optimizing performance**: Prioritize and assign resources to enhance application performance and cluster efficiency.

    **Quality of Service (QoS) classes**: Kubernetes uses these (Guaranteed, Burstable, and BestEffort) to decide how to allocate resources.

    **Assignment**: Match the right QoS class to each pod based on its importance and resource needs to ensure optimal performance.

5.  **Adapting to needs**: Implement dynamic scaling to continuously meet the changing demands of applications.

**The role of the cluster autoscaler**: It automatically adjusts the size of the Kubernetes cluster based on the needs of the pods and the availability of resources.

**Benefit**: This keeps the cluster balanced in terms of resource availability and cost efficiency.

6.  **Continuous improvement**: Regularly update and refine resource management strategies to keep pace with evolving applications and workloads.

    **Ongoing process**: Regularly monitor and adjust resource allocation settings as applications and workloads evolve.

    **Goal**: Maintain an efficient, cost-effective, and stable cluster.

## Optimizing container image size and management

Optimizing container image size and management revolves around creating and handling images in a way that maximizes efficiency in deployment and runtime environments. The size of container images significantly influences the deployment speed and resource utilization in Kubernetes clusters. Smaller images are quicker to pull from registries, require less storage, and can improve the overall performance of the system.

The process begins with choosing minimal base images. Base images that contain only the necessary components for an application reduce the overall image size. For example, using a lightweight base image such as Alpine Linux instead of a full-fledged Ubuntu or CentOS image can drastically decrease the size.

During the image build process, it's essential to eliminate unnecessary files and dependencies. This step involves removing temporary build files, extraneous build dependencies, and unused libraries before the final image is created. Such an approach not only trims down the image size but also bolsters security by minimizing the attack surface.

Utilizing multi-stage builds is a key strategy. This approach allows for one image to be used to build the application and a different, leaner image to run it. This means that the final image only contains the necessary components for running the application, leaving out build tools and intermediate artifacts.

Effective image versioning plays a critical role in image management. Implementing a systematic versioning strategy ensures the correct deployment of images and simplifies rollback procedures. Periodic cleanup of unused images from both development and production registries helps in efficient storage management and reduces clutter.

Layer caching is a technique that enhances build efficiency. By caching frequently used layers, build times are reduced, and network bandwidth is conserved. In scenarios where changes are made to certain layers while others remain unchanged, cached layers can be reused, speeding up the build process.

Integrating security scanning into the image build and deployment process is vital. Regular scans for vulnerabilities in container images help in identifying and mitigating security risks. Automated scanning tools can be integrated into the **continuous integration and continuous deployment (CI/CD)** pipeline for this purpose.

Optimizing the retrieval of images in a Kubernetes cluster is also important. Using a private container registry located close to the Kubernetes cluster can reduce image pull times. Implementing a judicious image pull policy in Kubernetes, such as `IfNotPresent`, can prevent unnecessary image downloads, conserve network resources, and expedite pod startup times.

To provide a clear context of optimizing a Dockerfile using multi-stage builds, let's consider a scenario where you are developing a simple Node.js application.

Multi-stage builds allow you to use separate stages to build the application and run it, resulting in a significantly smaller final image.

### Step 1 – define the base image for building

Start with a base image that includes all the necessary build tools. In this case, we'll use a Node.js image that includes the full Node.js runtime and npm package manager, both of which are needed to install dependencies and build the application:

```
Dockerfile
# Stage 1: Build the application
FROM node:16 as builder

# Set the working directory
WORKDIR /app

# Copy package.json and package-lock.json
COPY package.json package-lock.json ./

# Install dependencies
RUN npm install

# Copy the rest of your application code
COPY . .

# Build the application (if applicable)
RUN npm run build
```

### Step 2 – use a minimal base image for the runtime

After building the application, switch to a lighter base image for the runtime stage. Alpine Linux is a good choice due to its minimal size:

```
# Stage 2: Setup the runtime environment
FROM node:16-alpine

# Set the working directory
```

```
WORKDIR /app

# Copy only the built artifacts and necessary files from the builder
stage
COPY --from=builder /app/build ./build
COPY --from=builder /app/node_modules ./node_modules

# Set the command to start your app
CMD ["node", "build/app.js"]
```

Let's take a closer look at each step:

1. **Building stage**:

   - The FROM node:16 as builder statement starts the first stage (builder) using the Node.js 16 image

   - The application's dependencies are installed using npm install

   - All necessary build commands are run to compile the application or perform any tasks required to prepare the app for deployment

2. **Runtime stage**:

   - FROM node:16-alpine starts the second stage using a much smaller base image, Alpine Linux

   - Essential files from the build stage are copied over. The COPY --from=builder syntax indicates copying from the earlier build stage

   - Only the artifacts that are necessary to run the application are included in the final image, significantly reducing its size

Choosing base images and managing layers are critical decisions in containerization and significantly affect the performance, security, and maintainability of applications in a Kubernetes environment. Let's look at the key trade-offs and considerations in the decision-making process.

**Choosing base images**:

- **Size versus functionality**:

   I.   **Smaller images**: Opting for minimal base images such as Alpine Linux can drastically reduce image size, leading to faster pull times and reduced attack surface. However, minimal images may lack the necessary libraries or tools, which can complicate the setup.

   II.  **Fuller images**: Larger base images, such as those from Ubuntu or CentOS, might include many built-in utilities and libraries, simplifying development and debugging but increasing the image size and potentially introducing more security vulnerabilities.

- **Compatibility and stability**:

    I.    **Stable images**: More substantial and well-established distributions (for example, Ubuntu) are tested across a wide range of environments and are known for stability. This can be critical for complex applications.

    II.    **Edge cases**: Smaller or less common base images may offer advantages in terms of performance but can sometimes lead to compatibility issues with libraries or tools needed for your applications.

- **Security**:

    I.    **Vulnerabilities**: Larger images can contain more packages, which potentially increases the attack surface. Choosing images that are frequently updated and minimizing the installed packages are essential steps in maintaining security.

    II.    **Maintenance and updates**: It's vital to select base images from repositories that provide regular and reliable security updates to mitigate newly discovered vulnerabilities.

**Managing layers**:

- **Layer optimization**:

    I.    **Fewer layers**: Reducing the number of layers in your image can improve pull times and storage efficiency. Using multi-stage builds to separate build environments from runtime environments helps in minimizing the final image layers.

    II.    **Layer caching**: Thoughtful ordering of steps in Dockerfiles ensures that more stable commands (less likely to change) are at the top and more dynamic commands (more likely to change) are at the bottom. This strategy leverages Docker's caching mechanism effectively, reducing build times during development.

- **Reusability versus specificity**:

    I.    **Generic layers**: Common layers across multiple images, such as operating system base layers, can be reused, saving storage space and speeding up image pulls in environments where multiple containers use the same base.

    II.    **Custom layers**: Specific layers that address the unique needs of an application ensure that the container only includes what's necessary for operation, reducing size and potentially increasing security.

- **Build time versus runtime efficiency**:

    I.    **Build optimization**: While optimizing the number of layers can reduce build time and storage needs, sometimes, additional layers during the build phase (such as separating dependency installation from code copying in development) can speed up subsequent builds due to better use of cache

II.    **Runtime optimization**: Ensuring that the runtime image is as lean as possible typically means sacrificing some build-time efficiency for a smaller, more efficient runtime environment

**Decision-making process**: In the decision-making process, you should consider the following aspects:

- **Application requirements**: What does your application need in terms of libraries, tools, and runtime environments?

- **Security policies**: What are your organizational requirements for security? This might dictate certain base images over others.

- **Resource efficiency**: How critical are the size and speed of your container deployments?

- **Maintenance and support**: How well supported and maintained are the base images you are considering?

## Network performance tuning

To ensure that applications running in the cluster communicate efficiently and reliably, network performance tuning is a critical aspect. This involves optimizing various network components and settings within the Kubernetes environment.

The first area of focus is network plugins. Kubernetes supports different **container network interface** (**CNI**) plugins, and choosing the right one can significantly impact network performance. Some plugins are optimized for specific use cases, such as high throughput or low latency, and selecting one that aligns with the cluster's needs is vital.

Another key aspect is tuning network policies. Network policies in Kubernetes control how pods communicate with each other and with other network endpoints. Optimizing these policies helps in reducing unnecessary network traffic, improving security, and potentially enhancing network performance. It's important to define clear, concise rules that only allow the required traffic, reducing the overhead on the network.

Implementing service mesh technology can also contribute to network performance. A service mesh such as Istio or Linkerd provides advanced network features such as load balancing, fine-grained control, and monitoring, which are essential for managing complex microservices-based applications. These tools can optimize traffic flow and improve the reliability and efficiency of network communication.

Monitoring and analyzing network traffic is crucial for tuning. Tools such as Wireshark, tcpdump, or more Kubernetes-centric tools such as Cilium can be used to monitor network packets. This monitoring helps in identifying bottlenecks, abnormal traffic patterns, or issues such as packet loss and latency, which can then be addressed.

DNS performance is often overlooked but is crucial in Kubernetes. Optimizing DNS resolution times and ensuring the scalability of the DNS service within Kubernetes can greatly impact overall network efficiency. This might involve tuning the DNS configuration, using more efficient DNS servers, or optimizing caching.

Load balancing strategies within Kubernetes also play a significant role in network performance. Efficient load balancing ensures that no single node or pod is overwhelmed with traffic, leading to better response times and reduced latency. This might involve tuning the settings of Ingress controllers or load balancers used within the cluster.

Ensuring optimal TCP/IP settings on nodes can make a significant difference. Settings such as TCP window size, keep-alive settings, and others can be tuned based on the specific network characteristics and requirements of the cluster.

In addition to these, implementing network **quality of service** (**QoS**) and considering the physical network infrastructure (such as using high-bandwidth connections, ensuring proper routing, and more) are important. Network QoS ensures that critical traffic is prioritized, and having robust physical network infrastructure supports the overall network performance of the cluster.

By focusing on these areas, Kubernetes administrators can significantly enhance the network performance of their clusters, ensuring that applications are responsive, scalable, and reliable in their communication needs.

## Enhancing data storage performance

In a Kubernetes environment, the efficiency and speed of data-intensive applications depend significantly on optimizing the data storage layer. This process involves a combination of selecting suitable storage options, configuring persistent volumes, and implementing performance-enhancing strategies.

The choice of storage solution is critical. Kubernetes supports various types, such as block, file, and object storage, each with its strengths for different workload types. Factors that influence this choice include the application's performance needs, scalability requirements, and data persistence characteristics.

Configuring **persistent volumes** (**PVs**) and **persistent volume claims** (**PVCs**) is a key step. Adjusting the storage provisioner, access modes, and storage classes can lead to significant performance improvements. High-performance storage options, such as SSDs, are beneficial for I/O-intensive workloads.

Data caching mechanisms play a significant role in enhancing performance. By storing frequently accessed data in memory or on faster storage media, read/write operations become more efficient, particularly for applications with repetitive access patterns.

Tuning storage I/O is crucial for optimal data throughput and minimal latency. Adjusting parameters such as queue depth and buffer sizes to match the application's requirements can align storage performance with workload demands.

Advanced storage features such as snapshots and replication not only aid in data protection but also contribute to performance. Snapshots offer quick recovery options through point-in-time data copies, while replication ensures data availability and resilience.

Monitoring storage performance using tools such as Prometheus and Grafana is essential for maintaining optimal operation. These tools help identify usage patterns, bottlenecks, and areas needing improvement.

Network optimizations specific to storage can also yield performance gains. Employing high-speed networks for storage traffic and optimizing network protocols can reduce data transfer times and enhance efficiency.

Balancing storage capacity with performance requires a dynamic approach. Auto-scaling storage solutions that adjust resources based on current demand ensure that applications always have sufficient storage without resource wastage.

Keeping storage drivers and firmware updated is crucial for maintaining compatibility and performance in a Kubernetes environment. Regular updates prevent issues related to performance degradation and ensure smooth operation.

Overall, the focus on optimizing data storage in Kubernetes centers around carefully selecting and managing storage solutions, fine-tuning configurations and performance parameters, and a consistent monitoring and maintenance regime. This ensures that the storage infrastructure supports the diverse requirements of Kubernetes-based applications.

## Utilizing resource quotas and limits effectively

Effectively utilizing resource quotas and limits is a key strategy for managing the resources that are available in a cluster and preventing any single application or user from consuming more than their fair share. This management is crucial in multi-tenant environments where cluster resources are shared among different teams or projects.

Resource quotas are applied at the namespace level. They act as a ceiling for the total resources that can be consumed by all the pods within a given namespace. Quotas can encompass various resource types, including CPU, memory, and storage, as well as the count of resources, such as pods, services, and persistent volume claims. By setting quotas, administrators can control the impact of each namespace on the overall cluster, preventing any single namespace from overconsuming resources and affecting other operations.

At the pod or container level, resource limits define the maximum amount of CPU and memory that each container can use. Kubernetes enforces these limits to ensure that a container doesn't exceed its allocated share. If a container tries to use more CPU than its limit, Kubernetes throttles the CPU usage. If a container exceeds its memory limit, it might be terminated, a mechanism that protects other containers from being starved of resources.

Setting these quotas and limits requires a deep understanding of the applications' resource needs. This understanding is gained through monitoring and analysis. If set too low, quotas and limits can choke applications, causing performance issues or even outages. If set too high, they may lead to underutilization of resources. The goal is to strike a balance where resources are allocated fairly and efficiently, without overprovisioning or wastage.

Typically, Kubernetes administrators set resource quotas by creating a `ResourceQuota` object in a namespace. This object specifies the limits across various resource types. For example, it can limit the total amount of memory and CPU that all pods in the namespace can consume, or it can restrict the number of persistent volume claims that can be created.

`LimitRange` objects are used to set default resource requests and limits for pods and containers in a namespace. This ensures that every pod or container has some basic level of CPU and memory allocation, and it prevents any single pod from monopolizing resources. `LimitRange` objects also help in maintaining the QoS for pods, ensuring that critical applications get the resources they need.

Continuous monitoring is crucial in this context. Tools such as Kubernetes Metrics Server provide data on resource usage, helping administrators adjust quotas and limits in response to changing requirements and usage patterns. This monitoring and adjustment are ongoing tasks and are essential for maintaining the efficiency and stability of the Kubernetes environment.

Adjusting resource quotas and limits in Kubernetes based on real-time data and metrics involves a cyclical process of monitoring, analyzing, and adapting. Administrators use tools such as Prometheus and Kubernetes Metrics Server to continuously monitor resource usage. Based on these insights, they can dynamically adjust quotas and limits to optimize performance and resource utilization. Here are some key adjustments to consider:

- **Updating ResourceQuota objects**: Modifying CPU, memory, and storage limits at the namespace level

- **Tweaking LimitRange settings**: Setting default and maximum resource consumption per pod to ensure fair allocation

- **Using autoscalers**: Implementing Kubernetes autoscalers such as VPA and HPA to adjust resources automatically based on load and performance metrics

This ongoing adjustment ensures that resources are allocated efficiently, thereby maintaining cluster stability and preventing resource wastage or contention.

## Efficient logging and monitoring strategies

Establishing efficient logging and monitoring strategies in a Kubernetes environment plays a crucial role in maintaining the operational health and performance of applications and the cluster. These strategies enable activities to be tracked, anomalies to be detected, and issues to be resolved quickly.

Centralized logging is key in a distributed system such as Kubernetes. It involves aggregating logs from all components, including pods, nodes, and Kubernetes system components, into a central repository. Using an **Elasticsearch, Fluentd, Kibana** (**EFK**) stack or similar solutions such as Graylog helps in efficiently managing and analyzing logs from various sources. This centralized approach simplifies searching, filtering, and analyzing log data, making it easier to pinpoint issues.

Setting appropriate log levels is essential for effective logging. Log levels control the verbosity of the log messages. Fine-tuning these levels ensures that the logs capture necessary information without overwhelming the storage with irrelevant data. For instance, DEBUG or INFO levels might be suitable for development environments, while ERROR or WARN levels might be more appropriate in production.

System-level logs, including those from Kubernetes components such as the API server, scheduler, controller manager, kubelet, and container runtime, are vital for understanding the health and behavior of the cluster. Monitoring these logs provides insights into the Kubernetes system's operations and helps in identifying issues related to cluster management and orchestration.

On the monitoring front, collecting and analyzing metrics gives a quantitative view of the cluster's performance. Metrics such as CPU, memory usage, network I/O, and disk throughput are critical for assessing the health of both the cluster and the applications running on it. Application-specific metrics also provide valuable insights into the performance and behavior of individual applications.

Prometheus is a widely adopted tool in the Kubernetes ecosystem for monitoring. It scrapes metrics from multiple sources, stores them efficiently, and allows for complex queries and alerts. When integrated with Grafana, it offers a powerful visualization tool, enabling the creation of detailed dashboards that reflect the state of the Kubernetes cluster and applications.

Alerting mechanisms based on metric thresholds are fundamental to proactive monitoring. By setting up alerts, administrators can be notified of potential issues as they arise, allowing for timely intervention before they escalate into more significant problems.

Implementing proper health checks with liveness and readiness probes in Kubernetes helps maintain application reliability. Liveness probes detect and remedy failing containers, ensuring that applications are running correctly. Readiness probes determine when a container is ready to start accepting traffic, preventing requests from being routed to containers that are not fully operational yet.

For organizations, customizing log aggregation and analysis tools can be done by adjusting the complexity and scalability of the solution to match their specific needs.

Here's a guide presented in a simpler format:

| Consideration | Small Organizations | Medium Organizations | Large Organizations |
| --- | --- | --- | --- |
| Centralized Logging | Use lightweight open source solutions such as Loki or EFK with limited retention | Implement robust solutions such as EFK, with scalability for growing traffic | Deploy enterprise-level EFK stacks with high availability and long-term storage |
| Log Levels | Set to higher verbosity for in-depth monitoring due to fewer resources | Optimize log levels for a balance between detail and storage efficiency | Configure lower verbosity for production, focusing on errors and warnings to manage large volumes of logs |

| Consideration | Small Organizations | Medium Organizations | Large Organizations |
|---|---|---|---|
| System-Level Logs | Focus on critical component logs to reduce overhead | Monitor a broader set of components for deeper insights | Implement comprehensive logging across all components, possibly using a tiered storage solution |
| Metrics and Monitoring | Basic metrics collection with a simple dashboard for key insights | Advanced metrics collection with detailed dashboards for different user roles | Integrate with sophisticated monitoring solutions that provide predictive analytics and complex queries |
| Alerting Mechanisms | Simple alert rules based on critical thresholds | More complex alerts that incorporate trends and patterns | Highly customized alerts that integrate with incident management systems for automated responses |
| Health Checks | Implement the necessary liveness and readiness checks | Use advanced health checks with automated recovery solutions | Integrate health checks with auto-scaling and self-healing mechanisms for optimal performance |

## Load balancing and service discovery optimization

Optimizing load balancing and service discovery in Kubernetes is fundamental for ensuring efficient distribution of traffic across your applications and services. This optimization leads to improved application responsiveness and reliability.

Load balancing in Kubernetes is typically handled by services and Ingress controllers. Services provide an internal load balancing mechanism, distributing incoming requests to the right pods. Fine-tuning service specifications, such as choosing between `ClusterIP`, `NodePort`, and `LoadBalancer` types, depending on the use case, is crucial. For external traffic, Ingress controllers play a pivotal role. They manage external access to the services, typically through HTTP/HTTPS, and can be configured for more complex load balancing, SSL termination, and name-based virtual hosting.

Optimizing these Ingress controllers is vital. Selecting the right Ingress controller that aligns with your performance and routing requirements is essential. Configuration options such as setting up efficient load balancing algorithms (round-robin, least connections, IP hash, and so on) and tuning session affinity parameters can significantly affect performance and user experience.

Service discovery in Kubernetes allows pods to locate each other and communicate efficiently. It uses DNS for service discovery, where services are assigned DNS names and pods can resolve these names to IP addresses. Ensuring that the DNS system within Kubernetes is optimized is crucial for service discovery performance. This includes configuring the DNS cache properly to reduce DNS lookup times and managing DNS query traffic efficiently.

Implementing service mesh technologies such as Istio or Linkerd can further enhance load balancing and service discovery. Service meshes offer sophisticated traffic management capabilities that go beyond what's available with standard Kubernetes services and Ingress controllers. They can provide fine-grained control over traffic with features such as canary deployments, circuit breakers, and detailed metrics, which are invaluable for optimizing performance and reliability.

Another aspect to consider is the effective management of network policies. Network policies in Kubernetes control how pods communicate with each other and with other network endpoints. By defining precise network policies, you can ensure efficient traffic flow and enhance the security of your applications.

For high-availability scenarios, setting up multi-zone or multi-region load balancing is important. This ensures that traffic is distributed across different geographical locations, improving the application's resilience and providing a better experience for users spread across various regions.

Regularly monitoring the performance of your load balancing and service discovery mechanisms is also key. This involves tracking metrics such as request latency, error rates, and throughput, which helps in identifying bottlenecks and areas for improvement.

In practice, optimizing load balancing and service discovery in Kubernetes involves a combination of choosing the right tools and technologies, fine-tuning configurations, and continuous monitoring and adjustment. This approach ensures that traffic is efficiently distributed and services are easily discoverable, leading to enhanced performance and reliability of applications running in Kubernetes.

When optimizing Ingress controllers in Kubernetes, you will be configuring them for efficient load balancing and advanced traffic management.

Common Ingress controllers include NGINX and Traefik. Let's learn how to optimize them.

The following YAML provides an example of setting up an NGINX Ingress controller with session affinity and a least connections load balancing algorithm:

```
apiVersion: networking.k8s.io/v1
kind: Ingress
metadata:
  name: nginx-example-ingress
  annotations:
    nginx.ingress.kubernetes.io/load-balance: "least_conn" # Load
balancing algorithm
    nginx.ingress.kubernetes.io/affinity: "cookie" # Enable session
affinity
    nginx.ingress.kubernetes.io/session-cookie-name: "nginxaffinity" #
Set cookie name
    nginx.ingress.kubernetes.io/session-cookie-hash: "sha1" # Hash
algorithm for cookie
spec:
  ingressClassName: nginx
```

```
    tls:
    - hosts:
      - myapp.example.com
        secretName: myapp-tls-secret
    rules:
    - host: myapp.example.com
      http:
        paths:
        - path: /
          pathType: Prefix
          backend:
            service:
              name: myapp-service
              port:
                number: 80
```

Make sure you have the correct `ingressClassName` for your NGINX Ingress setup and the `myapp-tls-secret` TLS secret if you're using HTTPS.

Here's a full YAML for a Traefik IngressRoute with a round-robin load balancing strategy and middleware for basic auth:

```
apiVersion: traefik.containo.us/v1alpha1
kind: IngressRoute
metadata:
  name: traefik-example-ingressroute
spec:
  entryPoints:
    - web
  routes:
  - match: Host(`myapp.example.com`)
    kind: Rule
    services:
    - name: myapp-service
      port: 80
      strategy: RoundRobin
    middlewares:
    - name: auth-middleware


---
apiVersion: traefik.containo.us/v1alpha1
kind: Middleware
metadata:
```

```
    name: auth-middleware
  spec:
    basicAuth:
      secret: myapp-basic-auth-secret
```

The middleware for basic authentication refers to a Kubernetes secret called `myapp-basic-auth-secret`, which you need to create beforehand as it contains the encoded credentials.

**Applying the configuration**:

Save your chosen configuration to a file – for example, `nginx-ingress.yaml` or `traefik-ingress.yaml`.

Apply the configuration using `kubectl`:

```
kubectl apply -f nginx-ingress.yaml
```

Here's how you can do this for Traefik:

```
kubectl apply -f traefik-ingress.yaml
```

## Implementing proactive node health checks

Proactive node health checks in Kubernetes are crucial for early detection and resolution of issues, which, in turn, maintains the reliability and performance of the cluster. These checks focus on continuously monitoring the status and health of nodes, preempting potential problems that could impact the cluster.

Key to this approach is the use of Kubernetes' built-in functionalities, such as Node Condition and Node Problem Detector. Node Condition provides insights into various aspects of a node's status, including CPU, memory, disk usage, and network availability. By closely monitoring these conditions, administrators can quickly identify nodes that are facing resource constraints or operational issues.

Node Problem Detector augments these capabilities. It's designed to detect specific issues, such as kernel errors, hardware failures, and critical system service failures. By reporting these problems as node conditions, it brings attention to issues that might otherwise remain unnoticed until they cause significant disruption.

Integrating additional monitoring tools such as Prometheus into the Kubernetes environment offers a more comprehensive view of node health. Prometheus can collect a broad spectrum of metrics, allowing for detailed tracking of resource usage, system performance, and the operational health of each node. These metrics provide essential data points for identifying trends, diagnosing issues, and making informed decisions about resource management and capacity planning.

Automating response mechanisms is a significant part of proactive node health checks. Configuring automated actions for common scenarios, such as draining and restarting unresponsive nodes, ensures quick resolution of issues with minimal manual intervention. This automation can be further enhanced by integrating Kubernetes features such as Cluster Autoscaler, which automatically replaces nodes that are consistently failing health checks, maintaining the resilience and capacity of the cluster.

Regularly maintaining and updating the node infrastructure is vital for preventing issues. Keeping the operating system, Kubernetes components, and other critical software up-to-date helps avoid vulnerabilities and compatibility issues that could lead to node health problems.

Conducting periodic load tests on nodes is an effective way to proactively identify potential performance issues. These tests simulate high-load conditions, revealing how nodes behave under stress and highlighting areas where performance improvements may be needed.

Therefore, implementing proactive node health checks in Kubernetes involves a blend of utilizing built-in tools for monitoring, integrating advanced monitoring solutions, automating responses to detected issues, maintaining the node infrastructure, and conducting regular load testing. This comprehensive approach ensures that nodes remain healthy and capable of efficiently supporting the cluster's workloads.

With that, we have thoroughly examined various techniques to enhance the performance of Kubernetes environments, from fine-tuning resource allocation to optimizing network and data storage performance. These strategies are essential for maintaining a robust and responsive Kubernetes infrastructure, ensuring that each component operates at its peak efficiency.

Now, we will transition our focus toward ensuring that these systems are not only performing well but are also scalable and efficient in the long term. The upcoming section will delve into the architectural decisions and scaling strategies that can support sustainable growth and adaptability in your Kubernetes environment. From adopting microservices architecture to exploring cluster federation, we will explore how to design systems that can easily expand and adapt to changing demands without compromising performance.

# Ensuring efficiency and scalability

This section focuses on ways to boost efficiency and scalability in Kubernetes. It discusses stateless design, adopting microservices, cluster auto-scaling, and different scaling strategies.

## Designing for statelessness and scalability

Creating applications that are both stateless and scalable is a core principle in Kubernetes design that aims to enhance the efficiency and responsiveness of services. This approach involves structuring applications in a way that minimizes their reliance on internal state, which, in turn, facilitates easy scaling and management.

In stateless application design, each request to the application must contain all the information necessary to process it. This means the application doesn't rely on information from previous interactions or maintain a persistent state between requests. This design is inherently scalable as any instance of the application can handle any request, allowing for easy horizontal scaling.

A key benefit of statelessness is the simplicity it brings to scaling and load balancing. Since any instance of the application can respond to any request, Kubernetes can easily distribute traffic across multiple instances of the application without needing complex logic to maintain a session or user state.

Implementing a stateless architecture often involves moving state management out of the application. This can be achieved by using external data stores such as databases or caching services for maintaining session data, user profiles, or other transactional data. These external services must be scalable and highly available to ensure they don't become bottlenecks as the application scales.

Containerization inherently supports stateless design. Containers are ephemeral and can be easily started, stopped, or replaced. This aligns well with the principles of statelessness, where the loss of a single container does not impact the application's overall state or functionality.

In Kubernetes, Deployments and ReplicaSets are ideal for managing stateless applications. They ensure that a specified number of pod replicas are running at any given time, facilitating easy scaling up or down based on demand. **Horizontal Pod Autoscaler** (HPA) in Kubernetes can automatically scale the number of pod replicas based on observed CPU utilization or other select metrics, further enhancing the application's scalability.

Design patterns such as the 12-Factor App methodology provide guidelines that are beneficial for stateless application development. These patterns emphasize factors such as code base, dependencies, configuration, backing services, and processes, guiding developers in building applications that are optimized for cloud environments and scalability.

Load testing is an essential part of validating the scalability of stateless applications. Regularly testing the application under varying loads helps in understanding its behavior and limitations, allowing for informed decisions on infrastructure needs and scaling policies.

In designing stateless applications for scalability in Kubernetes, the focus is on ensuring that applications do not maintain an internal state and can handle requests independently. This approach simplifies the deployment, scaling, and management of applications, making them more robust and adaptable to changing loads and environments.

## Adopting microservices architecture appropriately

Adopting a microservices architecture in Kubernetes is about breaking down applications into smaller, independent services, each running in its own container. This approach offers numerous advantages for scalability and efficiency but requires careful planning and implementation to be successful.

Microservices enable individual components of an application to be scaled independently. Unlike monolithic architectures, where scaling often requires the entire application stack to be replicated, microservices can be scaled based on their specific needs. This leads to more efficient resource utilization and can address bottlenecks more effectively.

The agility in development and deployment is a significant benefit of microservices. Teams can focus on specific areas of an application, leading to faster development cycles, easier testing, and quicker deployment. This modular approach also facilitates more frequent updates and rapid iteration of individual components without impacting the entire application.

Kubernetes provides a robust environment for orchestrating microservices. Its ability to manage containerized applications lends itself well to a microservice architecture, handling complex tasks such as service deployment, scaling, load balancing, and self-healing of services.

However, microservices introduce complexities, particularly in service-to-service communication. Ensuring efficient and secure communication between microservices is crucial. Kubernetes offers tools for service discovery and networking to facilitate this communication, but these require thoughtful configuration to optimize performance and maintain security.

Data management is another critical aspect of a microservices setup. Ideally, each microservice manages its own data, which helps in maintaining the independence of services. However, this leads to challenges in ensuring data consistency and managing transactions across different services.

In a microservices environment, centralized logging and monitoring become even more important. With multiple independent services, it's essential to have a unified view of the system's health and performance to quickly identify and address issues in any of the microservices.

Security considerations are amplified in a microservices architecture. Each service introduces potential vulnerabilities, making it essential to implement strong security practices. Kubernetes network policies and secure service communication mechanisms are vital in safeguarding the microservices.

Transitioning to a microservices architecture from a monolithic setup should be approached incrementally. Starting with a single function or module and progressively expanding allows teams to adapt and learn the best practices for managing microservices in Kubernetes.

Microservices architecture in Kubernetes capitalizes on its strengths in managing dispersed, containerized services. This approach facilitates scalable, efficient application development but requires a strategic approach to service communication, data management, monitoring, and security.

## Cluster autoscaling techniques

Implementing cluster autoscaling techniques in Kubernetes is crucial for managing the dynamic resource requirements of applications efficiently. Autoscaling ensures that the cluster adjusts its size automatically based on workload demands, adding or removing nodes as necessary.

Cluster Autoscaler is a key component for achieving autoscaling. It monitors the resource usage of pods and nodes and automatically adjusts the size of the cluster. When it detects that pods cannot be scheduled due to resource constraints, it triggers the addition of new nodes. Conversely, if nodes are underutilized and certain conditions are met, it can remove nodes to reduce costs and improve efficiency.

The effectiveness of cluster autoscaling heavily depends on the right configuration and tuning of parameters. It involves setting appropriate thresholds for scaling up and down. These thresholds are based on metrics such as CPU utilization, memory usage, and custom metrics that reflect the workload's specific needs.

One important aspect is predicting and handling demand spikes. The autoscaler should be configured to respond quickly to increased demand to ensure that applications have the resources they need. However, it should also avoid overly aggressive scaling that can lead to unnecessary costs.

Integrating HPA with Cluster Autoscaler enhances these auto-scaling capabilities. While HPA adjusts the number of pod replicas within a node based on resource utilization, Cluster Autoscaler adjusts the number of nodes. Together, they ensure both efficient pod distribution and optimal node count.

**Pod disruption budgets** (**PDBs**) are crucial in maintaining application availability during scaling operations. They prevent the autoscaler from evicting too many pods from a node at once, which could lead to service outages.

In multi-tenant environments, balancing the needs of different applications and teams is a challenge for auto-scaling. Implementing namespace-specific resource quotas and priorities can help ensure fair resource allocation among different tenants when the cluster scales.

Cost management is another aspect of cluster autoscaling. While scaling up ensures that applications have the necessary resources, scaling down during periods of low demand can significantly reduce cloud infrastructure costs.

Regularly monitoring and analyzing autoscaling events and patterns is important. This data can provide insights for further tuning and optimization of the autoscaling parameters.

In essence, effectively implementing cluster autoscaling techniques requires a careful balance between responsiveness to workload demands and cost-efficiency. It involves configuring the autoscaler with appropriate thresholds, integrating it with pod-level scaling mechanisms, considering application availability, and regularly reviewing scaling patterns for ongoing optimization.

## Horizontal versus vertical scaling strategies

Understanding and choosing between horizontal and vertical scaling strategies is crucial for optimizing the performance and resource utilization of applications. These two strategies offer different approaches to handling increased workload demands, and selecting the right one depends on the specific needs of the application and the underlying infrastructure.

Horizontal scaling, also known as scaling out or in, involves adding or removing instances of pods to match the workload demand. This strategy is well-suited for stateless applications, where each instance can operate independently. Kubernetes facilitates horizontal scaling through ReplicaSets and Deployments, which allow for the easy addition or removal of pod instances. HPA can automate this process by adjusting the number of pod replicas based on observed CPU utilization or other select metrics.

The key advantage of horizontal scaling is that it can provide high availability and resilience. By distributing the load across multiple instances, it reduces the risk of a single point of failure. This approach also allows for more granular scaling as resources can be added incrementally, closely matching the demand.

On the other hand, vertical scaling, also known as scaling up or down, refers to adding or removing resources to or from an existing instance. In a Kubernetes context, this means increasing or decreasing the CPU and memory that's allocated to a pod. Vertical scaling is typically used for stateful applications or those that are difficult to partition into multiple instances.

Vertical scaling can be simpler to implement as it doesn't require the architectural considerations of horizontal scaling. However, it has its limitations. There are upper limits to how much a single instance can be scaled, and increasing resources often requires a restart of the pod, which can lead to downtime. Furthermore, vertical scaling doesn't address the issue of a single point of failure.

Deciding between horizontal and vertical scaling involves considering the nature of the application. Stateless applications, such as web servers, are generally good candidates for horizontal scaling due to their ability to run multiple instances simultaneously without conflict. Stateful applications, such as databases, might benefit more from vertical scaling, as they often rely on a single instance maintaining its state.

Another consideration is the cost and resource availability. Horizontal scaling can be more cost-effective in cloud environments where resources are billed based on usage, as additional instances can be added or removed to match the demand precisely. Vertical scaling, while simpler, might lead to underutilization or overutilization of resources as adjusting the size of instances is less granular.

In practice, a combination of both strategies might be employed in a Kubernetes environment. Some components of an application might scale horizontally, while others scale vertically, depending on their specific requirements and characteristics. This blended approach allows for both flexibility and efficiency in managing application scaling.

## Utilizing cluster federation for scalability

Utilizing cluster federation in Kubernetes is a strategy to enhance scalability and manage multiple Kubernetes clusters as a single entity. This approach is particularly useful in scenarios where applications are deployed across different regions, cloud providers, or data centers.

Cluster federation involves linking several Kubernetes clusters together, allowing for coordinated management of resources, services, and applications across these clusters. This setup enables central control while maintaining the autonomy of individual clusters. It's especially beneficial for organizations that require high availability, global distribution, and cross-region disaster recovery.

The primary advantage of cluster federation is the ability to spread workloads across multiple clusters and regions. This distribution can significantly improve application performance by bringing services closer to users, reducing latency, and ensuring compliance with data sovereignty requirements.

Additionally, it provides a mechanism for failover, where workloads can be shifted from one cluster to another in case of a failure or maintenance in a particular region.

In a federated setup, deploying and managing applications across different clusters becomes more streamlined. You can deploy an application to multiple clusters simultaneously, ensuring consistency in configuration and deployment. This approach simplifies the complexity involved in managing deployments across multiple environments.

Resource sharing and balancing across clusters is another aspect of federation. It allows for more efficient use of resources by moving workloads to clusters with spare capacity. This capability ensures that no single cluster is overburdened while others are underutilized.

DNS-based global load balancing can be integrated with cluster federation. This involves using a global DNS service that routes user requests to the nearest or best-performing cluster. Such a setup improves user experience by reducing response times and increasing service reliability.

However, cluster federation in Kubernetes also introduces complexity. Managing multiple clusters requires careful planning and robust infrastructure. There is a need for strong network connectivity between clusters, and security becomes a more prominent concern when data is distributed across multiple environments.

Managing stateful applications in a federated environment can be challenging. Data replication and consistency across geographically distributed clusters need to be handled with precision to avoid data conflicts and ensure reliability.

Monitoring and logging in a federated environment require a comprehensive approach. Centralized monitoring and logging solutions are essential to maintain visibility into the health and performance of applications and infrastructure across all federated clusters.

Overall, utilizing cluster federation in Kubernetes offers significant benefits for scalability, high availability, and global distribution. It enables efficient management of multi-cluster environments, optimizes resource utilization, and improves application performance. However, it requires careful implementation and you must consider aspects such as network connectivity, security, data management, and centralized monitoring.

## Efficient resource segmentation with namespaces

Efficient resource segmentation with namespaces is a strategy for organizing and managing resources within a cluster. Namespaces provide a way to divide cluster resources between multiple users, teams, or projects, enabling more efficient and secure management of the Kubernetes environment.

Namespaces act as virtual clusters within a single physical Kubernetes cluster. They allow for the isolation of resources, enabling different teams or projects to work within the same cluster without interfering with each other. Each namespace can contain its own set of resources, including pods, services, replication controllers, and deployments, making it easier to manage permissions and quotas.

One of the primary benefits of using namespaces is the ability to implement resource quotas and limits. Administrators can assign specific resource quotas to each namespace, controlling the maximum amount of CPU, memory, and storage that can be consumed by the resources within that namespace. This prevents any single team or project from consuming more than its fair share of cluster resources, ensuring fair allocation and preventing resource contention.

Namespaces also enhance security and access control within a Kubernetes cluster. **Role-based access control** (**RBAC**) can be used in conjunction with namespaces to grant users or groups specific permissions within their assigned namespaces. This fine-grained access control helps maintain security and operational integrity as users can only manage resources within their designated namespaces.

Organizing resources into namespaces simplifies the management and tracking of costs associated with running applications in Kubernetes. By associating specific namespaces with different teams or projects, it becomes easier to monitor and report on resource usage and allocate costs accordingly.

Namespaces also play a role in service discovery within Kubernetes. Services within the same namespace can discover each other using short names, which simplifies communication between microservices that are logically grouped. However, if needed, services in different namespaces can still communicate with each other using fully qualified domain names.

In multi-tenant environments, namespaces are essential for isolating different tenants' workloads. This isolation is crucial not only for resource management and billing but also for ensuring privacy and security between different tenants' applications.

Implementing namespaces requires careful planning and consideration of the overall cluster architecture. Decisions about how to divide resources, assign quotas, and configure access controls need to align with the organizational structure and requirements.

Efficient resource segmentation with namespaces is a powerful way to manage and allocate resources effectively. It supports multi-tenancy, enhances security, simplifies resource management, and aids in cost allocation. However, it requires thoughtful implementation to ensure that the namespaces are structured and managed in a way that aligns with the needs and goals of the organization.

## Optimizing inter-pod communication

To achieve efficient and reliable interactions between services in a Kubernetes cluster, it's crucial to optimize the communication between pods. This optimization is a key factor in enhancing the performance and scalability of containerized applications.

Central to this is the configuration of Kubernetes services, which provide a stable and abstract way to expose applications running in pods. By properly setting up services such as `ClusterIP`, `NodePort`, or `LoadBalancer`, administrators can define how pods communicate within the cluster and with external entities, impacting the overall efficiency of inter-pod interactions.

Implementing network policies is vital for managing traffic flow between pods. These policies allow administrators to specify exactly which pods can communicate with each other, enhancing security by limiting connections to only those that are necessary and authorized. This targeted approach to communication not only bolsters security but also streamlines network traffic.

Efficient service discovery and DNS configuration are also key components. Kubernetes automatically assigns DNS names to services, simplifying the process by which pods locate and communicate with each other. Ensuring that the cluster's DNS service is correctly configured and performing optimally is essential for seamless service discovery.

Advanced load balancing techniques play a significant role in evenly distributing network traffic across multiple pods, preventing any single pod from becoming overwhelmed. This can be achieved through Kubernetes Ingress controllers or service mesh solutions such as Istio or Linkerd, which offer sophisticated traffic management capabilities, including SSL termination and path-based routing.

Monitoring the network performance between pods is another important aspect. Utilizing tools such as Prometheus for metric collection and Grafana for data visualization, administrators can track and analyze network latency, throughput, and error rates. This ongoing monitoring enables the identification and resolution of communication bottlenecks or inefficiencies.

The choice of CNI plugin and network drivers impacts container network performance. Selecting the most suitable CNI plugin for the specific needs of the applications can lead to more efficient packet processing and reduced communication latency.

In scenarios with diverse and heavy network traffic, implementing network QoS can help prioritize critical or sensitive traffic. This ensures that high-priority communications are maintained even under heavy load conditions.

Application design also influences the efficiency of inter-pod communication. Avoiding overly frequent or complex interactions between microservices and designing services to be as autonomous as possible can significantly reduce the overhead and complexity of communication within the Kubernetes environment.

Through these strategies – configuring services and network policies, optimizing DNS and load balancing, monitoring network performance, selecting appropriate network interfaces, and adhering to best practices in application design – Kubernetes administrators can effectively optimize inter-pod communication. This optimization is key to ensuring not just the speed and efficiency of communications but also their reliability and security within the Kubernetes cluster.

## Load testing and capacity planning

Load testing and capacity planning are integral components of managing a Kubernetes environment and are crucial for ensuring that applications can handle expected traffic volumes and the cluster has sufficient resources to meet demand.

Load testing involves simulating real-world traffic to an application to assess how it performs under various conditions. This process is vital for identifying potential bottlenecks and issues that might not be apparent under normal usage. By gradually increasing the load on the application and monitoring its performance, administrators can determine the maximum capacity it can handle before it starts to degrade in terms of response time or reliability.

In a Kubernetes context, load testing should cover not just the application but also the underlying infrastructure, including pod scalability, database performance, and networking capabilities. Tools such as Apache JMeter, Locust, or custom scripts can generate the required load on applications. Monitoring tools such as Prometheus, coupled with Grafana for visualization, are used to track key performance metrics during the tests.

The results of load testing inform capacity planning, which is the process of predicting future resource requirements to handle anticipated load increases. Capacity planning in Kubernetes involves determining the appropriate number and size of nodes, the right scaling policies for pods, and ensuring adequate network and storage resources.

Effective capacity planning requires a thorough understanding of both the current resource utilization and the expected growth in traffic and application complexity. It often involves analyzing historical usage data and traffic patterns to forecast future needs. This data can be used to create models that predict how additional load will affect the system.

Autoscaling strategies in Kubernetes, both HPA and Cluster Autoscaling, play a critical role in capacity planning. These strategies allow the cluster to automatically adjust the number of running pod replicas and nodes based on the current load, ensuring that the application has the resources it needs while minimizing unnecessary resource usage.

Considering peak traffic times is important in capacity planning. The system should be capable of handling sudden spikes in traffic without performance degradation. This often involves over-provisioning resources to some extent to accommodate unexpected surges in demand.

Capacity planning also involves considering the trade-offs between cost and performance. While it's important to have enough resources to handle peak loads, over-provisioning can lead to unnecessary expenses. Finding the right balance is key to efficient resource utilization.

Regularly revisiting and updating the capacity plan is essential as application requirements and traffic patterns can change over time. Continuous monitoring, regular load testing, and analysis of traffic trends help maintain an up-to-date understanding of capacity needs.

It's an ongoing process that helps ensure applications are robust and responsive. It involves testing applications under realistic load scenarios, analyzing performance data, predicting future resource requirements, and continuously adjusting resource allocation to meet changing demands efficiently and cost-effectively.

Having explored the key strategies for ensuring efficiency and scalability in Kubernetes, we've covered everything from the fundamental design principles to advanced scaling techniques. These insights are crucial for creating Kubernetes environments that are not only robust but also capable of growing and adapting efficiently as demands increase.

Next, we will shift our focus to maximizing the full potential of Kubernetes. The upcoming section will explore a range of powerful features and integrations that extend Kubernetes capabilities. From harnessing its extensibility with custom resources to adopting sophisticated deployment and management strategies such as GitOps, we will uncover how to leverage Kubernetes in more dynamic, versatile, and effective ways to meet modern IT and business demands.

# Maximizing the potential of Kubernetes

This section addresses maximizing Kubernetes' potential by exploring its extensibility with custom resources, integration with cloud-native ecosystems, continuous deployment, advanced scheduling, container runtime optimization, data management, hybrid and multi-cloud strategies, and the adoption of GitOps for effective Kubernetes management.

## Harnessing Kubernetes extensibility with custom resources

Kubernetes' extensibility through custom resources is a powerful feature that allows developers to add new functionalities and resources to the Kubernetes API. This capability enables the creation of declarative APIs that are as easy to use as built-in Kubernetes resources.

**Custom resource definitions** (**CRDs**) are used to define custom resources in Kubernetes. They provide a way for developers to define the structure of their own resources and API objects. Once a CRD is created, users can create and access instances of this new resource with `kubectl`, just like they would with built-in resources such as Pods and Services.

The use of custom resources opens up a world of possibilities for extending Kubernetes' functionalities. They allow new types of services, applications, and frameworks to be integrated into the Kubernetes ecosystem, making the platform more adaptable to specific needs and use cases.

Operators are a key pattern in leveraging custom resources. An Operator is a method of packaging, deploying, and managing a Kubernetes application. It builds upon custom resources and custom controllers. Operators use the Kubernetes API to manage resources and handle the operational logic, automating complex tasks that are typically done by human operators.

Custom controllers are another aspect of Kubernetes' extensibility. They watch for changes to specific resources and then trigger actions in response. When combined with custom resources, custom controllers can manage entire life cycles of services, from deployment to scaling to monitoring.

The implementation of custom resources and controllers can enhance automation within Kubernetes. For example, a custom resource could be created to manage a database cluster, with a custom controller that handles backups, scaling, and updates automatically based on the specifications defined in the custom resource.

Security is a vital consideration when extending Kubernetes with custom resources. It's important to ensure that custom resources and controllers are designed with security in mind, following best practices such as least privilege and regular auditing.

Extending Kubernetes with custom resources also involves careful consideration of cluster performance and stability. Custom controllers should be designed to be efficient and responsive, avoiding excessive API calls that could overwhelm the Kubernetes API server.

Kubernetes' extensibility with custom resources enables the creation of tailored solutions that fit specific operational needs. By defining new resource types and automating their management with custom controllers and operators, developers can significantly enhance the functionality and efficiency of their Kubernetes environments. This extensibility makes Kubernetes a versatile platform that can adapt to a wide range of applications and workflows.

## Integrating with cloud-native ecosystems

The integration of Kubernetes with cloud-native ecosystems is a vital step in leveraging the full potential of modern infrastructure and services. Kubernetes, being a cornerstone of the cloud-native landscape, is designed to work seamlessly with a variety of tools and platforms that adhere to cloud-native principles.

Cloud-native ecosystems are composed of various tools and technologies that work together to provide a comprehensive environment for building, deploying, and managing containerized applications. These ecosystems typically include CI/CD tools, monitoring and logging solutions, service meshes, and cloud-native storage systems.

Integrating CI/CD pipelines with Kubernetes is essential for automating the deployment process. Tools such as Jenkins, GitLab CI, and Spinnaker can be used to build, test, and deploy applications automatically to Kubernetes, making the process faster, more reliable, and less prone to human error.

Monitoring and logging are crucial components of the cloud-native ecosystem. Tools such as Prometheus for monitoring and the EFK stack for logging provide insights into the health and performance of applications running in Kubernetes. These tools can be integrated into the Kubernetes environment to collect metrics and logs, enabling real-time monitoring and efficient troubleshooting.

Service meshes such as Istio, Linkerd, and Consul add an additional layer of control and observability to Kubernetes. They provide advanced networking features, such as traffic management, security, and observability, without requiring changes to the application code. Integrating a service mesh into a Kubernetes environment can greatly simplify the management of inter-service communications and enhance the overall security and reliability of the applications.

Cloud-native storage solutions are another critical aspect of integration. As Kubernetes applications often require persistent storage, integrating with cloud-native storage solutions such as Ceph, Rook, or Portworx ensures that applications have scalable, reliable, and performant storage available to them.

Incorporating security tools and practices into the Kubernetes environment is also important. Integrating security tools such as Aqua Security, Twistlock, or Sysdig can help in continuously scanning for vulnerabilities, enforcing security policies, and ensuring compliance with security standards.

The integration process also involves adapting Kubernetes applications to be cloud-agnostic, ensuring that they can run on any cloud platform without significant changes. This is particularly important for organizations that operate in multi-cloud or hybrid cloud environments.

Automation plays a key role in managing the Kubernetes ecosystem. Tools such as Terraform or Ansible can be used for automating the deployment and management of Kubernetes clusters and the associated cloud-native infrastructure.

Integrating Kubernetes with cloud-native ecosystems requires a strategic approach that combines selecting the right tools and technologies, configuring them to work together seamlessly, and continuously monitoring and optimizing their performance. This integration is key to building a robust, scalable, and efficient Kubernetes environment that fully leverages the benefits of cloud-native technologies.

## Leveraging Kubernetes for continuous deployment

The implementation of continuous deployment within Kubernetes environments transforms how organizations approach software releases, making the process faster and more reliable. Kubernetes provides a range of features that streamline and automate the deployment pipeline, allowing for more frequent and consistent updates to applications.

At the heart of leveraging Kubernetes for continuous deployment is the integration of robust CI/CD pipelines. Tools such as Jenkins, GitLab CI, or CircleCI can be set up to automatically build, test, and deploy code changes to Kubernetes, creating a seamless flow from code commit to deployment.

Kubernetes facilitates continuous deployment through its declarative configuration and automated management of application states. Developers specify the desired state of applications using manifest files, and Kubernetes automatically applies these changes, maintaining the system's state as defined.

Rolling updates are a cornerstone of Kubernetes' deployment capabilities, ensuring that new application versions are released with minimal disruption. This approach incrementally updates application instances, which helps maintain service availability and reduces the risk of introducing errors.

For more controlled deployments, Kubernetes supports advanced strategies such as canary and blue-green deployments. Canary deployments allow for new versions to be rolled out to a limited audience first, while blue-green deployments involve running two identical environments with different versions, providing an option to switch over once the new version is verified.

Autoscaling capabilities in Kubernetes align well with continuous deployment practices. The platform can dynamically adjust the number of running instances based on the current load, ensuring optimal performance even as new versions are rolled out.

Effective monitoring and logging, enabled by Kubernetes-compatible tools such as Prometheus for performance metrics and the EFK stack for logging, are vital. They provide visibility into the application's performance and help quickly pinpoint issues in new releases.

Kubernetes namespaces offer a way to segregate environments within the same cluster, such as development, staging, and production. This separation helps manage deployments across different stages of development without risk to the production environments.

In the event of deployment issues, Kubernetes facilitates automated rollbacks. This feature quickly reverts the application to its previous stable version, minimizing the impact of any deployment-related problems.

By harnessing these features, Kubernetes becomes an enabler of continuous deployment, allowing development teams to release updates more frequently and with greater confidence. The platform's ability to automate deployment processes, manage application states, and ensure high availability makes it an ideal choice for organizations looking to embrace a more agile and responsive software delivery approach.

## Utilizing advanced scheduling features

Kubernetes offers advanced scheduling features that enable more precise and efficient placement of pods on nodes in the cluster. These features allow administrators and developers to control how pods are scheduled, taking into account the specific needs of the workloads and the characteristics of the cluster nodes.

One of the key advanced scheduling features in Kubernetes is node affinity and anti-affinity. Node affinity allows you to specify rules for pod placement based on node attributes. For example, you can ensure that certain pods are placed on nodes with specific hardware such as SSDs or GPUs, or in a particular geographic location. Node anti-affinity, on the other hand, ensures that pods are not co-located on the same node, which is crucial for high-availability setups and load balancing.

Pod affinity and anti-affinity extend these capabilities to the pod level. They allow you to define rules for pod placement relative to other pods. For example, you can configure pods to be scheduled on the same node as other pods from the same or different services, which can be useful for reducing latency or ensuring that related components are co-located.

Taints and tolerations are other powerful scheduling features. Taints are applied to nodes and mark them as unsuitable for certain pods, while tolerations are applied to pods and allow them to be scheduled on nodes with specific taints. This mechanism is useful for dedicating nodes to specific types of workloads or for keeping certain workloads off specific nodes.

Pod priority and preemption enable Kubernetes to schedule pods based on priority levels. Pods with higher priority can be scheduled before lower-priority pods and, if necessary, trigger the preemption of lower-priority pods to free up resources on nodes. This feature is essential for ensuring that critical workloads get the resources they need.

Resource quotas and limit ranges are also crucial in advanced scheduling. They allow administrators to manage the consumption of cluster resources such as CPU and memory more effectively. By setting quotas and limits at the namespace level, you can control resource allocation among multiple teams or projects, ensuring fair usage and preventing resource starvation.

The Kubernetes scheduler can also be extended with custom schedulers. This allows for the creation of custom scheduling logic that can address unique requirements or optimize scheduling for specific types of workloads, such as data-intensive applications or microservices with particular interdependencies.

DaemonSets ensure that a copy of a specific pod runs on all or some nodes in the cluster. This is particularly useful for running pods that provide system services such as log collectors or monitoring agents on every node.

To effectively utilize advanced scheduling features in Kubernetes, it's important to understand the specific requirements of your applications and the available resources in your cluster. These features provide the flexibility to optimize pod placement for performance, availability, and resource utilization, ensuring that the Kubernetes cluster operates efficiently and effectively.

## Container runtime optimization

Optimizing the container runtime in Kubernetes is essential for enhancing the overall performance and efficiency of containerized applications. The container runtime is responsible for managing the life cycle of containers within a Kubernetes cluster, including their creation, execution, and termination.

Selecting the right container runtime can have a significant impact on performance. Kubernetes supports several runtimes, including Docker, containerd, and CRI-O. Each runtime has its own set of features and performance characteristics, and the choice depends on specific workload requirements, security considerations, and compatibility with existing systems.

Efficient image management is a key aspect of runtime optimization. This involves using smaller and more efficient container images to reduce startup times and save bandwidth. Multi-stage builds in Docker, for example, can help in creating leaner images by separating the build environment from the runtime environment.

Optimizing resource allocation to containers is crucial for runtime performance. This includes setting appropriate CPU and memory requests and limits for each container. Properly configured resource limits ensure that containers have enough resources to perform optimally while preventing them from monopolizing system resources.

Runtime security is also an important consideration. Securing the container runtime involves implementing security best practices such as using trusted base images, regularly scanning images for vulnerabilities, and enforcing runtime security policies using tools such as AppArmor, seccomp, or SELinux.

Network performance optimization is another aspect of container runtime optimization. This involves configuring network plugins and settings for optimal throughput and latency. Kubernetes offers various CNI plugins, each with different networking features and performance profiles.

Storage performance optimization is vital, especially for I/O-intensive applications. This includes selecting the appropriate storage drivers and configuring storage options to balance performance and reliability. Persistent storage solutions in Kubernetes should be chosen based on their performance characteristics and compatibility with the container runtime.

Monitoring and logging are essential for identifying and addressing runtime performance issues. Tools such as Prometheus for monitoring and Fluentd or Logstash for logging can provide insights into the runtime's performance, helping to detect and troubleshoot issues.

Regular updates and maintenance of the container runtime and its components are important for performance and security. Keeping the runtime and its dependencies up-to-date ensures that you benefit from the latest performance improvements and security patches.

So, optimizing the container runtime in Kubernetes involves selecting the right runtime, efficiently managing container images, allocating resources appropriately, ensuring security, optimizing network and storage performance, implementing effective monitoring and logging, and regularly maintaining and updating the runtime environment. These steps are crucial for maximizing the performance and efficiency of containerized applications in a Kubernetes cluster.

## Effective data management and backup strategies

Ensuring the integrity, availability, and durability of data within a Kubernetes cluster relies heavily on the meticulous planning and execution of data storage, backup, and recovery solutions tailored to the requirements of containerized applications.

For data storage, Kubernetes supports a variety of persistent storage options, such as PVs and PVCs, which can be backed by different storage solutions, such as cloud storage, **network-attached storage** (**NAS**), or block storage systems. Choosing the right storage solution is vital for balancing performance, scalability, and cost. Factors such as I/O performance, data volume size, and access patterns should guide the selection process.

Implementing dynamic provisioning of storage using Storage Classes in Kubernetes simplifies the management of storage resources. Storage Classes allow administrators to define different types of storage with specific characteristics, and PVCs can automatically provision the required type of storage on demand.

Backup strategies in Kubernetes should be comprehensive, covering not only the data but also the cluster configuration and state. Regular backups of application data, Kubernetes objects, and configurations ensure that you can quickly recover from data loss or corruption scenarios.

The choice of backup tools and solutions should consider the specific requirements of Kubernetes environments. Solutions such as Velero, Stash, and Kasten K10 are designed to handle the complexities of Kubernetes backup and recovery, including backing up entire namespaces, applications, and persistent volumes.

For stateful applications, such as databases, implementing application-consistent backups is important. This ensures that the backups capture a consistent state of the application, including in-flight transactions. Techniques such as snapshotting and write-ahead logging can be employed to achieve application-consistent backups.

Disaster recovery planning is an extension of the backup strategy. It involves not only regularly backing up data but also ensuring that the backups can be restored in a different environment. This might involve cross-region or cross-cloud backups, enabling recovery even in the case of a complete regional outage.

Regularly testing backup and recovery processes is critical. Frequent testing ensures that backups are being performed correctly and that data can be reliably restored within the expected timeframes. This testing should be part of the regular operational procedures.

Data encryption, both at rest and in transit, is a key aspect of data management in Kubernetes. Encrypting data protects it from unauthorized access and ensures compliance with regulatory requirements. Kubernetes supports encryption at various levels, including storage-level encryption and network encryption for data in transit.

Automating data management and backup processes through Kubernetes' native features or third-party tools can significantly reduce the risk of human error and ensure consistent application of policies.

Implementing effective data management and backup strategies in Kubernetes requires a combination of the right storage solutions, comprehensive backup and recovery plans, regular testing, data encryption, and automation. These components work together to safeguard data against loss or corruption and ensure that applications running in Kubernetes can reliably and securely manage their data.

## Hybrid and multi-cloud deployment strategies

Deploying applications in a hybrid or multi-cloud environment is an increasingly popular strategy in Kubernetes as it offers flexibility, resilience, and optimization of resources. This approach allows organizations to leverage the strengths of different cloud environments and on-premises infrastructure, catering to a diverse set of operational requirements and business needs.

In a hybrid cloud setup, Kubernetes clusters are distributed across on-premises data centers and public clouds. This arrangement combines the security and control of private infrastructure with the scalability and innovation of public cloud services. It's ideal for organizations that have legacy systems on-premises but want to take advantage of cloud capabilities.

Multi-cloud deployments involve running Kubernetes clusters on different public cloud platforms. This strategy avoids vendor lock-in, provides high availability across different geographical locations, and allows organizations to use specific cloud services that best meet their application requirements.

A key component of successful hybrid and multi-cloud deployments is a consistent and unified management layer. Tools such as Rancher, Google Anthos, and Azure Arc enable centralized management of multiple Kubernetes clusters, regardless of where they are hosted. These tools simplify operations by providing a single pane of glass for deploying applications, monitoring performance, and enforcing security policies across all environments.

Networking is a critical aspect of hybrid and multi-cloud strategies. Ensuring reliable and secure communication between clusters in different environments can be challenging. Implementing network overlays or using cloud-native network services can provide seamless connectivity. Additionally, service meshes such as Istio and Linkerd can manage inter-cluster communication, providing consistent traffic management and security policies across clouds.

Data management and storage strategies must also be adapted for hybrid and multi-cloud environments. Considerations include data locality, compliance with data sovereignty laws, and ensuring high availability and disaster recovery across cloud boundaries. Using cloud-agnostic storage solutions or **container storage interfaces** (**CSIs**) can provide consistent storage experiences across different clouds.

Workload portability is another important factor. Containers inherently support portability, but it's crucial to design applications and their dependencies to be cloud-agnostic. This might involve using containerized microservices, abstracting cloud-specific services, or using APIs that are compatible across different cloud providers.

Security and compliance are heightened concerns in hybrid and multi-cloud environments. Implementing robust security practices, such as identity and access management, network security policies, and regular security audits, is essential. Compliance with various regulatory standards may also require specific controls and measures in different cloud environments.

Cost management and optimization are challenging but essential in hybrid and multi-cloud deployments. Tools and practices for monitoring and optimizing cloud expenses are vital to ensure that resources are used efficiently and costs are controlled.

Adopting hybrid and multi-cloud deployment strategies with Kubernetes offers significant benefits in terms of flexibility, scalability, and resilience. However, it also introduces complexities related to management, networking, data storage, portability, security, and cost. Careful planning and the use of appropriate tools and practices are essential for navigating these challenges and fully realizing the advantages of these deployment models.

# Adopting GitOps for Kubernetes management

The GitOps methodology revolutionizes Kubernetes management by applying the familiar principles of Git – version control, collaboration, and CI/CD automation – to infrastructure and deployment processes. This approach centers around using Git as the foundational tool for managing and maintaining the desired state of Kubernetes clusters.

In a GitOps workflow, the entire state of the Kubernetes cluster, including configurations and environment definitions, is stored in Git repositories. Changes to the cluster are made by updating the manifests or configuration files in these repositories. This method ensures that all changes are traceable, auditable, and subject to version control, just like any code changes.

Tools such as Argo CD, Flux, and Jenkins X play a crucial role in automating the synchronization between the Git repository and the Kubernetes cluster. These tools continuously monitor the repository for changes and apply them to the cluster, ensuring that the actual state of the cluster matches the desired state defined in Git.

One of the most significant advantages of adopting GitOps is the enhancement of deployment reliability. Automating deployments through Git merge requests or pull requests creates a consistent, repeatable, and error-resistant process. This streamlined approach significantly reduces the likelihood of errors that can occur with manual deployments.

GitOps also fosters better collaboration among team members. Since all changes are made through Git, they can be reviewed, commented on, and approved collaboratively. This openness not only improves the quality of changes but also facilitates knowledge-sharing and transparency within the team.

The version control aspect of GitOps provides a detailed audit trail of all changes made to the Kubernetes environment. Teams can easily track who made what changes and when, which is invaluable for maintaining security and compliance standards. In case of any issues, teams can quickly revert to a previous state, enhancing the resilience of the system.

By codifying everything, GitOps inherently promotes better security practices. It encourages a shift-left approach, where security and compliance checks are integrated early in the deployment process, reducing the chances of vulnerabilities in the production environment.

Monitoring and alerting are integral to the GitOps approach. Since the desired state is declared and stored in Git, any drift from this state in the live environment can be detected and rectified automatically. This constant monitoring ensures the stability and consistency of the Kubernetes environment.

For teams embarking on the GitOps journey, a comprehensive understanding of Git workflows, Kubernetes manifests, and CI/CD processes is essential. Adequate training and skill development in these areas are crucial for a smooth transition to this methodology.

# Summary

This chapter covered a wide range of performance optimization techniques for Kubernetes, offering insights into effective resource management, container optimization, and network tuning. It discussed the critical aspects of data storage, resource quotas, logging, monitoring, and advanced strategies for load balancing and node health checks. The narrative also touched upon the scalability of Kubernetes, exploring stateless architectures, microservices, cluster scaling, and balancing horizontal and vertical scaling strategies. Additionally, this chapter discussed Kubernetes' potential for integration with cloud-native ecosystems, highlighting continuous deployment, advanced scheduling, container runtime optimization, and effective data management. It underscored Kubernetes' adaptability to various operational needs, emphasizing its role as a versatile platform for enhancing system operations and efficiency.

In the next chapter, we will explore the concept of continuous improvement in Kubernetes, discover its importance, and learn how to apply iterative practices while adapting to the evolving Kubernetes ecosystem for sustained excellence.

# Part 3: Achieving Continuous Improvement

In this part, you will grasp the concept of continuous improvement in Kubernetes, enabling you to optimize performance and efficiency across the entire Kubernetes environment. The focus is on instilling a mindset of perpetual growth and adaptation to maintain and enhance Kubernetes deployments amidst evolving challenges and opportunities.

This part contains the following chapters:

- *Chapter 7, Embracing Continuous Improvement in Kubernetes*

- *Chapter 8, Proactive Assessment and Prevention*

- *Chapter 9, Bringing It All Together*

# 7

# Embracing Continuous Improvement in Kubernetes

This chapter focuses on embracing continuous improvement in Kubernetes, a key strategy for keeping up with the fast-paced evolution of technology. It addresses various topics, from foundational concepts of continuous improvement to integrating feedback effectively in iterative processes. The chapter also contrasts traditional approaches with modern continuous improvement methods, discusses how to measure success in such initiatives, and highlights the psychological aspects of fostering a growth mindset. It also covers practical aspects such as continuous learning, aligning improvement with DevOps practices, and managing risks iteratively. Moreover, the chapter offers a guide on adapting to changes in the Kubernetes ecosystem, including adopting new features and updates and understanding the role of community and collaboration.

We'll cover the following topics in this chapter:

- The concept of continuous improvement
- Implementing iterative practices
- Adapting to the evolving Kubernetes ecosystem

# The concept of continuous improvement

This section explores the fundamentals of continuous improvement in Kubernetes, emphasizing the role of feedback loops, comparing it to traditional models, measuring success in initiatives, understanding the psychological aspect of a growth mindset, continuous learning, and its impact on team dynamics. Additionally, it discusses integrating continuous improvement with DevOps practices.

## Fundamentals of continuous improvement in Kubernetes

Understanding the fundamentals of continuous improvement in Kubernetes begins with recognizing the platform's ever-changing nature. Kubernetes is not a static tool; it evolves with the technological landscape, responding to new needs and challenges. This characteristic demands a mindset geared toward ongoing refinement and enhancement.

Central to this approach is keeping up with Kubernetes' updates. These updates can include new features, security enhancements, and performance improvements. Staying informed about these changes is critical. It ensures the Kubernetes environment remains effective and up to date. Teams need to commit to continuous learning, ensuring they're aware of the latest developments and how they can be applied for better performance and efficiency.

Regularly reviewing and assessing the Kubernetes setup is another key step. This process should cover all aspects of Kubernetes, from how clusters are configured to deployment strategies. Such reviews help identify improvement areas, whether it's in efficiency, scalability, security, or maintainability.

Experimentation is also vital. Kubernetes' flexibility allows for trying out different configurations and approaches. Finding more effective ways to use Kubernetes often comes from this willingness to experiment. However, it's important to ensure new methods are thoroughly tested before they are used in more critical environments.

Feedback is a crucial element. Collecting and analyzing data from the system through monitoring and logging and from users through surveys or direct communication offers insights that guide improvements. It ensures that the Kubernetes environment aligns technically and meets user needs.

Automating routine tasks is a significant step toward continuous improvement. Automation in Kubernetes can range from simple scripts to sophisticated **continuous integration and continuous deployment (CI/CD)** pipelines. It reduces human error and frees up time for teams to focus on strategic tasks.

Collaboration and sharing knowledge are fundamental too. Kubernetes environments often involve different teams and stakeholders. Promoting open communication and collaboration fosters a comprehensive approach to managing and improving Kubernetes.

Setting measurable goals and metrics is important for tracking progress. These should align with the objectives of the Kubernetes environment, such as reducing deployment times or improving system reliability.

Risk management is also a key component. Anticipating and mitigating potential risks ensures that improvements do not compromise the system's stability or security.

Lastly, cultivating a culture of resilience and adaptability helps teams respond effectively to challenges and changes. Teams that adapt well to change are more likely to integrate continuous improvement into their workflow, leading to a stronger and more effective Kubernetes environment.

These fundamentals form the backbone of a continuous improvement strategy in Kubernetes, emphasizing the need for an adaptive, informed, and collaborative approach.

## The role of feedback loops in Kubernetes' evolution

Feedback loops are essential in the evolution of Kubernetes environments. They offer a structured approach to gathering and analyzing information, which is key in identifying areas for improvement. In Kubernetes, feedback can come from various sources, such as system logs, monitoring tools, and user feedback. Each of these sources provides valuable insights into how the Kubernetes environment is performing and how it can be improved.

System logs in Kubernetes offer a wealth of information. They record events and actions taken by the system, which can be used to track down issues and understand how changes in configuration affect the system's performance. By regularly reviewing these logs, teams can spot patterns and anomalies that might indicate potential problems or areas for optimization.

Monitoring tools are another critical component of feedback loops. These tools provide real-time data on the health and performance of Kubernetes clusters. This data helps teams to quickly identify and respond to issues such as resource bottlenecks or failing services. Moreover, monitoring tools can be configured to alert teams to specific conditions, enabling them to react swiftly to maintain system stability and performance.

User feedback is equally important in the Kubernetes evolution process. Users of the Kubernetes environment, whether they are internal development teams or external clients, can provide insights that are not immediately apparent from system logs or monitoring tools. This feedback can cover a wide range of aspects, from the ease of deploying applications to the performance of services running on Kubernetes. Actively seeking and incorporating this feedback ensures that the Kubernetes environment aligns with user needs and expectations.

Implementing effective feedback loops in Kubernetes requires a systematic approach. This involves setting up the necessary tools and processes to collect feedback, analyzing this feedback to extract meaningful insights, and then using these insights to guide improvements in the Kubernetes environment. It's a continuous process that helps keep the Kubernetes system aligned with evolving requirements and industry best practices.

Feedback loops encourage a proactive approach to managing Kubernetes environments. Instead of reacting to problems after they occur, teams can use feedback to anticipate and prevent issues. This proactive stance not only improves the reliability and performance of the Kubernetes system but also enhances the overall experience for those who depend on it.

Feedback loops are crucial in managing Kubernetes environments effectively, but they can encounter several pitfalls and obstacles. Here's a brief discussion on common challenges and strategies to overcome them.

### Common pitfalls in feedback loop management

Pitfalls frequently encountered in feedback loop management include the following:

- Overwhelming data
- Feedback silos
- Delayed responses
- Lack of actionable insights

### Strategies to overcome these obstacles

The following strategies are recommended:

- Implement tools and processes that can filter and prioritize data automatically, focusing on the most relevant information to manage noise and prevent information overload.

- Ensure that feedback from all sources is collected in a centralized system where it can be correlated and analyzed collectively.

- Use monitoring tools configured with automated alerts to respond quickly to critical issues, reducing the time between problem identification and resolution.

- Establish a culture of continuous improvement where feedback is regularly analyzed for insights and the findings are quickly implemented to refine Kubernetes operations.

## Comparing continuous improvement to traditional models

Comparing continuous improvement in Kubernetes to traditional models reveals a shift in mindset and approach toward managing IT infrastructure and applications. Traditional models often rely on a more static, linear progression of development and deployment. These models typically involve long planning phases, followed by implementation and a final review stage. Changes are infrequent and usually require a complete cycle to implement new ideas or address issues.

In contrast, continuous improvement in Kubernetes embraces a more dynamic and iterative approach. This method is characterized by frequent, incremental changes rather than large-scale overhauls. In the Kubernetes context, this means continuously updating and refining configurations, deployments, and the cluster itself to respond to new requirements or to improve efficiency and reliability.

One of the key differences lies in how feedback is integrated. In traditional models, feedback is often gathered at the end of a development cycle, which can delay the implementation of essential changes. With continuous improvement, feedback is an ongoing process, integrated into every stage of development and deployment. This immediate integration of feedback allows for quicker adaptation, enhancing the agility of the system and the team managing it.

Another significant difference is in the area of risk management. Traditional models often view changes as potential risks that need to be minimized, leading to a cautious approach toward updates and improvements. Continuous improvement in Kubernetes, however, perceives change as an opportunity for enhancement. While risks are still carefully managed, there is a greater willingness to experiment and iterate, leading to a more resilient and adaptable system.

The role of automation is considerably more pronounced in continuous improvement. Traditional models may utilize automation, but in the Kubernetes ecosystem, automation is a cornerstone of the continuous improvement process. It enables rapid deployment, consistent application of configurations, and instant rollback if needed, which are essential for maintaining a dynamic and responsive environment.

In terms of team dynamics and collaboration, continuous improvement encourages a more integrated and cross-functional approach. Traditional models often have distinct phases handled by separate teams, such as development, testing, and operations. Kubernetes, on the other hand, promotes a more collaborative environment where teams work together throughout the entire process, breaking down silos and enhancing communication.

Moreover, the approach toward learning and development differs significantly. Traditional models often rely on established practices and resist deviation from these norms. In contrast, continuous improvement in Kubernetes fosters a culture of ongoing learning and adaptation, where new tools, techniques, and practices are continually explored and integrated.

This comparison shows that continuous improvement in Kubernetes is not just about implementing a set of tools or practices. It represents a fundamental shift in how organizations approach the development, deployment, and management of applications and infrastructure. This shift enables more responsive, efficient, and effective management of Kubernetes environments, better aligning with the fast-paced and ever-changing nature of modern technology landscapes:

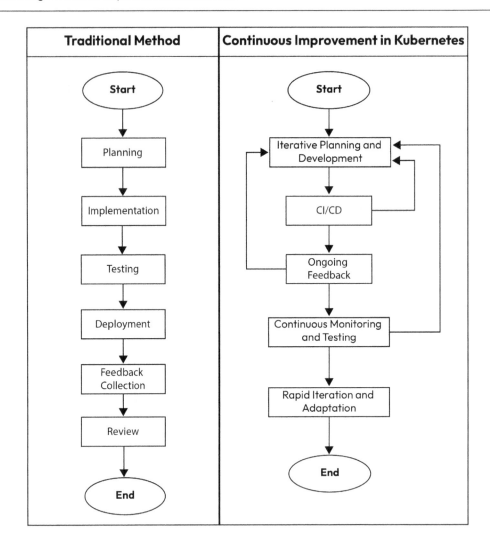

Figure 7.1 – Traditional model versus continuous improvement in Kubernetes

## Measuring success in continuous improvement initiatives

Measuring success in continuous improvement initiatives within Kubernetes environments requires a multi-faceted approach. Success isn't just about the immediate results; it also involves the long-term sustainability and adaptability of the Kubernetes system. To effectively gauge success, several **key performance indicators** (**KPIs**) and metrics are essential.

To begin with, deployment frequency serves as a primary indicator. Frequent and successful deployments suggest a healthy, continuously improving Kubernetes environment. This metric reflects not only the team's ability to introduce changes but also the stability and reliability of the system.

Another crucial metric is the lead time for changes. This measures the time taken from committing a change to it being successfully running in production. Shorter lead times indicate a more efficient and responsive Kubernetes environment.

Error rates also provide valuable insights. Monitoring the number and severity of errors post-deployment can indicate the quality of the continuous improvement processes. A decrease in error rates over time suggests that the team is effectively learning from past mistakes and improving their practices.

System downtime and availability are equally important. **High availability** (**HA**) and minimal downtime are key goals in Kubernetes environments. Tracking these metrics helps assess the impact of continuous improvement efforts on system reliability.

Customer satisfaction is a metric that should not be overlooked. Feedback from end users provides direct insight into the effectiveness of the Kubernetes environment and the applications it supports. High satisfaction levels indicate that the system is meeting or exceeding user expectations.

Resource utilization efficiency is another critical factor. Effective continuous improvement initiatives often lead to better utilization of resources, reducing costs and improving overall system performance.

The pace of innovation can also be a measure of success. A Kubernetes environment that continuously evolves and adopts new features or technologies demonstrates a successful continuous improvement culture.

Team morale and engagement are somewhat intangible but incredibly significant. A motivated and engaged team is more likely to contribute effectively to continuous improvement efforts, leading to better outcomes.

The response to failure and the time taken to recover from it also serve as important indicators. A successful continuous improvement process enables teams to quickly identify, address, and recover from failures, minimizing their impact.

Aligning Kubernetes KPIs with broader business goals is essential to ensure that technical improvements directly support organizational objectives. This alignment can be facilitated by a structured framework or model that guides the integration of business strategies with technical performance metrics. Here's a step-by-step approach to achieving this alignment.

1. **Identify business objectives**

   **Objective**: Understand the primary goals of the organization, such as increasing market share, reducing costs, enhancing customer satisfaction, or speeding up product delivery.

   **Action**: Conduct meetings with key stakeholders to clarify these goals and how they relate to the Kubernetes environment.

2.  **Define relevant KPIs**

    **Objective**: Select KPIs that directly influence or reflect business objectives.

    **Action**: For each business objective, identify technical metrics in the Kubernetes environment that contribute to achieving these goals.

    **Examples**:

    **Increase market share**: Focus on deployment frequency and innovation pace to ensure rapid market responsiveness.

    **Reduce costs**: Track resource utilization and system efficiency to optimize expenditures.

3.  **Set specific targets**

    **Objective**: Establish clear, measurable targets for each KPI that reflect desired business outcomes.

    **Action**: Define quantitative goals for each KPI, such as "reduce deployment lead time by 30% within 6 months" or "achieve 99.9% system availability."

4.  **Integrate KPIs into continuous improvement processes**

    **Objective**: Ensure that KPIs are continuously monitored and that insights gained are fed back into the improvement loop.

    **Action**: Use monitoring tools to track these KPIs in real time and set up alerts for deviations from expected values. Incorporate regular reviews of these metrics into the continuous improvement cycle.

5.  **Communicate and collaborate**

    **Objective**: Maintain transparency and ensure that all team members understand how their actions contribute to business goals.

    **Action**: Regularly share KPI progress and challenges in cross-departmental meetings, ensuring technical teams and **business units** (**BUs**) are aligned.

6.  **Review and adjust**

    **Objective**: Adapt strategies based on feedback and changing business environments.

    **Action**: Conduct periodic strategic reviews to assess if the KPIs still align with business goals and make adjustments as necessary. This includes refining KPIs, setting new targets, or even redefining business objectives.

7.  **Celebrate success and learn from failures**

    **Objective**: Build a culture that values both success and constructive failure as opportunities for learning and development.

    **Action**: Recognize achievements that significantly impact business objectives and analyze shortfalls to understand their causes and improve future efforts.

# The psychological aspect of cultivating a growth mindset

Cultivating a growth mindset within Kubernetes environments plays a crucial role in both the personal development of team members and the overall success of projects. This mindset, characterized by an emphasis on learning, adaptability, and resilience, is particularly important in the fast-paced and ever-evolving landscape of Kubernetes and cloud-native technologies.

Teams that embrace a growth mindset in Kubernetes view challenges as opportunities for learning and development rather than as roadblocks. This perspective is vital for navigating the complexities and continual changes inherent in Kubernetes. It allows teams to approach problems with a solution-oriented mindset, fostering creativity and innovation.

This mindset also enhances the ability to adapt to change. Kubernetes, by its nature, is a dynamic platform that frequently evolves through updates and new features. Teams with a growth-oriented approach are more prepared to integrate these changes positively, viewing them as chances to improve both the system and their skill sets.

Collaboration and open communication are further enhanced by a growth mindset. In an environment as complex as Kubernetes, the sharing of knowledge and experiences is key to effective problem-solving. Teams that encourage learning from one another create a more inclusive and innovative working environment.

A significant benefit of the growth mindset is the constructive use of feedback. Continuous feedback, both from the Kubernetes system and its users, is a cornerstone of improvement. Teams that view this feedback as a learning opportunity can make more informed decisions and refine their strategies more effectively.

Continuous learning is another aspect closely tied to this mindset. The landscape of Kubernetes is in constant flux, with new tools and practices continually emerging. An attitude geared toward continuous learning ensures that team members remain up to date and skilled in the latest technological advancements.

Proactive problem-solving is also a feature of the growth mindset. Instead of reacting to issues as they arise, teams anticipate potential challenges and opportunities for improvement. This proactive approach often results in a more robust and efficient Kubernetes environment.

Innovation is driven by a willingness to experiment and take calculated risks. Teams that are open to exploring new methodologies and tools within Kubernetes can discover more efficient and effective ways of working, pushing the boundaries of what's possible in their environment.

Emphasizing personal and professional development complements the technical aspects of working with Kubernetes. Encouraging team members to broaden their skill sets, both directly and indirectly related to Kubernetes, fosters a more versatile and competent team.

Celebrating successes and learning from setbacks are also integral to this mindset. Recognizing and valuing achievements no matter the scale builds confidence and motivation. Similarly, viewing failures as learning experiences rather than setbacks contributes to a positive and forward-looking team atmosphere.

Incorporating the growth mindset into Kubernetes practices not only enhances the technical aspects of the environment but also builds a more resilient, adaptable, and innovative team culture. This psychological dimension is as important as technical skills in navigating the complex and ever-evolving world of Kubernetes.

## Continuous learning

Upgrading skills and knowledge is an essential component of working effectively with Kubernetes. This concept revolves around constantly upgrading skills and knowledge to stay abreast of the latest developments in this rapidly evolving technology. In the context of Kubernetes, continuous learning is not just about keeping up with new versions or features; it's about deepening an understanding of the entire ecosystem and improving the ways in which it is used.

In the Kubernetes landscape, technology and best practices evolve at a rapid pace. Professionals who dedicate themselves to continuous learning are better equipped to leverage new tools and methodologies as they emerge. This ongoing educational process ensures that teams can utilize the full capabilities of Kubernetes, leading to more efficient, secure, and robust deployments.

One of the key aspects of continuous learning in Kubernetes is staying updated with the latest releases and updates. Kubernetes is regularly updated with enhancements, security patches, and new features. Understanding these updates and integrating them into existing systems is crucial for maintaining a state-of-the-art environment.

Another important element is exploring the wider Kubernetes ecosystem, which includes related tools and services. This exploration enhances one's ability to build more comprehensive and effective solutions. It involves not only learning about direct Kubernetes-related technologies but also about surrounding tools that can optimize and complement Kubernetes deployments.

Hands-on experience is vital in the learning process. Practitioners often find that they gain deeper insights and a more practical understanding of Kubernetes by actively working with the system. This hands-on approach allows for experimentation and firsthand learning from both successes and challenges.

Community involvement is another avenue for continuous learning. Engaging with the Kubernetes community through forums, social media, conferences, and meetups provides exposure to a wealth of knowledge and experience. It's an opportunity to learn from others' experiences, share knowledge, and stay informed about emerging trends and best practices.

Professional training and certification programs are also beneficial. These programs provide structured learning paths and validate skills through recognized certifications. They are a way to ensure that the knowledge gained is comprehensive and in line with industry standards.

Self-study and research play a crucial role as well. With a plethora of resources available online, including official documentation, blogs, tutorials, and courses, individuals have access to a wide range of learning materials. This self-directed learning allows individuals to tailor their educational journey to their specific interests and needs.

Peer learning and knowledge sharing within teams are equally important. Teams that encourage sharing insights and experiences foster a collaborative learning environment. This collective approach to learning helps disseminate knowledge across the team, ensuring that everyone stays on the same page and can contribute effectively.

Reflecting on past experiences and projects is a valuable learning tool. By analyzing what worked well and what could be improved, individuals and teams can glean insights that guide future strategies and actions. This reflective practice is a key component of a mature learning process.

Continuous learning is not just a recommendation; it's a necessity. It empowers individuals and teams to keep pace with technological advancements, enhances their ability to solve complex problems, and ultimately leads to more successful and innovative Kubernetes deployments.

## The impact of continuous improvement on team dynamics

Continuous improvement in Kubernetes environments significantly influences team dynamics, nurturing a culture of collaboration, innovation, and mutual growth. This impact is observed in various aspects of team interactions and overall performance.

One of the primary effects is the enhancement of collaboration. Continuous improvement necessitates frequent communication and the sharing of ideas and solutions. As teams work together to identify areas for improvement, they develop a deeper understanding of each other's strengths and skills, leading to more effective teamwork and a stronger sense of unity.

This process also promotes a culture of shared responsibility. In a Kubernetes environment, where changes are constant and rapid, the traditional silos of roles become less defined. Developers, operations teams, and system administrators often find themselves working more closely, blurring the lines between their respective duties. This shared responsibility ensures that everyone feels invested in the project's success, nurturing a more cohesive and motivated team.

Innovation is another area where continuous improvement impacts team dynamics. The constant pursuit of better solutions and practices in Kubernetes encourages team members to think creatively and propose innovative ideas. This environment, where experimentation and calculated risk-taking are encouraged, leads to a more dynamic and forward-thinking team.

The focus on continuous improvement also facilitates personal and professional growth among team members. As the team strives to enhance the Kubernetes environment, individuals are encouraged to upgrade their skills and knowledge. This not only benefits the project but also contributes to each team member's career development, creating a more skilled and confident team.

Continuous improvement strengthens problem-solving skills. As teams regularly encounter and address challenges in the Kubernetes environment, they develop a more refined approach to problem-solving. This experience is invaluable as it equips team members with the ability to tackle complex issues more efficiently and effectively.

Team morale and motivation are positively affected as well. Achieving incremental improvements and seeing tangible results of their efforts gives team members a sense of accomplishment and purpose. This boosts morale and fosters a positive work environment where individuals feel valued and motivated.

Continuous improvement leads to more effective conflict resolution. As team members collaborate closely, they learn to communicate more effectively and resolve disagreements constructively. This improved communication is crucial in maintaining a harmonious and productive team dynamic.

The approach also encourages adaptability and flexibility among team members. In a constantly evolving Kubernetes environment, teams need to be able to quickly adjust to new tools, practices, and challenges. Continuous improvement cultivates this adaptability, making the team more resilient and capable of handling change.

Another impact is the nurturing of a supportive environment. As teams work together toward common goals, they build a supportive network where members help each other overcome challenges and share knowledge. This sense of support is vital for maintaining high levels of engagement and job satisfaction.

It's worth remembering that the emphasis on continuous improvement in Kubernetes environments brings about significant positive changes in team dynamics. It leads to enhanced collaboration, shared responsibility, innovation, personal growth, and a more resilient and supportive team culture. These changes not only benefit the project but also contribute to a more fulfilling and productive work environment for all team members.

At the same time, conflicts can arise more frequently due to stress, misunderstandings, or differing opinions on the direction of a project.

Potential conflicts in rapid change environments include the following:

- Role ambiguity
- Resource allocation
- Resistance to change
- Decision-making

These techniques may help to mitigate conflicts better:

- Regularly scheduled meetings, clear and open lines of communication, and established channels for feedback.
- Clearly define and regularly update roles and responsibilities for all team members.
- Involve the team in setting goals and objectives that accommodate rapid changes.
- Provide ongoing training and support to help team members adapt to new tools and practices.
- Employ a more democratic or participative approach in decision-making processes.
- Recognize and reward team members who adapt well to changes or who contribute positively during transitions.

# Integrating continuous improvement with DevOps practices

The fusion of continuous improvement and DevOps practices within Kubernetes environments is a strategic approach that significantly enhances the efficiency and effectiveness of software development and operations. This synergy capitalizes on the strengths of both methodologies, fostering an environment of ongoing enhancement and optimization.

Automation is a crucial element in this integration. DevOps already places a strong emphasis on automating repetitive tasks, and when combined with continuous improvement, this extends to identifying new areas for automation. Such practices not only streamline workflows in Kubernetes but also free up teams to focus on innovation and tackle more complex challenges.

Feedback loops are greatly enhanced in this integrated approach. Unlike traditional models where feedback might be delayed until post-deployment, continuous improvement intertwined with DevOps ensures immediate feedback. This immediacy allows for the rapid incorporation of insights into subsequent iterations, thereby accelerating improvements and refining the end product.

A culture of experimentation and learning is central to this approach. DevOps encourages testing new ideas, and continuous improvement provides a structured framework for these experiments. This environment enables teams to iterate quickly, learn from both successes and failures, and continuously refine their processes and tools.

Collaboration between development and operations teams sees a significant boost. The combination of continuous improvement with DevOps breaks down barriers of traditional silos, creating a more cohesive and integrated team environment. This collaborative approach is vital for comprehensive and effective improvements in both development and operational aspects.

Optimizing resource use is another key advantage of this integration. Efficient resource management, a core component of DevOps, is further enhanced by continuous improvement strategies. This leads to cost savings and improved performance in Kubernetes environments.

Risk management becomes more proactive in this context. Teams are better equipped to foresee and mitigate potential risks early on, safeguarding the stability and security of their Kubernetes environments.

Goal setting and metric tracking become more focused and aligned with organizational objectives. Clear, measurable targets for continuous improvement efforts ensure that they contribute effectively to the broader goals of the organization.

Scalability is also more effectively managed within this integrated framework. As Kubernetes environments grow in complexity, the blend of DevOps and continuous improvement practices ensures that scaling up systems and processes is efficient and minimally disruptive.

The integration of continuous improvement with DevOps practices within Kubernetes environments creates a dynamic and resilient framework. It leads to enhanced agility in software development and operations, higher-quality outcomes, and a robust, adaptable IT infrastructure that can efficiently evolve with organizational needs.

We've discussed the concept of continuous improvement in Kubernetes, exploring everything from the fundamentals to psychological and team dynamics aspects. This comprehensive approach highlights how continuous improvement is not just a set of practices but a transformative mindset that drives the evolution and effectiveness of Kubernetes environments.

Next, we will explore the implementation of iterative practices, a key component of continuous improvement. That deals with the principles of iterative development, structuring effective cycles, and learning from real-world case studies. By focusing on balancing speed and stability and integrating robust feedback mechanisms, we will uncover strategies to enhance the agility and responsiveness of Kubernetes deployments, ensuring they can adapt swiftly and efficiently to new challenges and opportunities.

# Implementing iterative practices

This section focuses on principles of iterative development in Kubernetes, effective cycle structuring, case studies, speed-stability balance, supportive tools, planning, feedback integration, and risk management strategies.

## Principles of iterative development in Kubernetes

Adopting an iterative development approach is key to effective system management and evolution. This method, characterized by gradual and continuous changes, aligns perfectly with the dynamic nature of container orchestration.

Starting with a minimal viable configuration and progressively building upon it is a fundamental aspect of this approach. In Kubernetes, this translates to implementing the most essential features first and then incrementally adding more complex functionalities. This strategy allows for testing and validation at each step, minimizing potential disruptions.

Frequent, small updates rather than large, infrequent ones are another crucial aspect. This tactic ensures that changes are manageable and any issues can be quickly identified and resolved. It contributes to a more stable and reliable Kubernetes environment, facilitating smoother updates and maintenance.

Collaboration across different teams is vital in iterative development. Developers, operations staff, and other stakeholders need to communicate continuously to maintain a shared understanding of the system's objectives and challenges. This collaboration is essential for swift decision-making and effective problem-solving.

Regular feedback, both from users and system performance data, is integral to refining Kubernetes configurations and applications. This continuous loop of feedback allows teams to adjust their strategies based on real-world usage and performance, ensuring the system meets user needs effectively.

Continuous testing and integration play a central role in this development style. With each iteration, it's important to ensure that new additions meet quality standards and integrate seamlessly with existing components. Utilizing automated testing and CI tools is crucial in this context.

Being adaptable is key in iterative development. Teams should be ready to alter their plans and strategies in response to new insights, technical challenges, or shifting requirements. This flexibility keeps the development process moving forward and ensures the Kubernetes environment remains relevant and efficient.

Simplicity and maintainability should be prioritized in design and configuration. A simpler, more maintainable Kubernetes setup reduces the risk of complications and makes scaling and management more straightforward.

Reflecting and evaluating regularly help drive continuous improvement. After each iteration, assessing what worked and what could be better sets the stage for ongoing refinement, ensuring each cycle brings valuable learning and improvement.

A user-centric focus is essential. Iterative development should always take into account the end user's needs and experiences to ensure that the Kubernetes environment serves its intended purpose effectively.

Setting clear, measurable goals for each iteration is important for tracking progress and maintaining focus. These goals act as benchmarks for success and help align team efforts with the broader objectives of the Kubernetes project.

By embracing these aspects of iterative development, teams can manage and evolve their Kubernetes environments more effectively, ensuring robustness, scalability, and alignment with organizational and user needs.

## Structuring effective iterative cycles

Effective iterative cycles hinge on establishing a well-defined process that allows for continuous improvement and adaptation. The aim is to develop, test, and deploy changes in a manner that maximizes efficiency and minimizes disruption.

Clear planning is foundational for effective iterative cycles. It entails setting specific, achievable objectives for each cycle, ensuring alignment with the broader goals of the Kubernetes project. These clear objectives help focus the team's efforts and provide a roadmap for the cycle's progression.

A key component is the establishment of short, manageable timeframes for each iteration. These timeframes should be long enough to achieve meaningful progress but short enough to maintain momentum and flexibility. This balance ensures that the team can quickly respond to feedback and changing requirements.

Incorporating regular checkpoints for review and assessment is vital. These checkpoints provide opportunities to evaluate progress against the set goals, identify any issues or challenges, and make necessary adjustments. Regular reviews help keep the team on track and ensure that the cycle is moving in the right direction.

Effective iterative cycles also require a strong emphasis on communication. Keeping all team members informed and engaged throughout the cycle is crucial for collaboration and ensuring everyone is aligned with the cycle's objectives and progress.

Another important aspect is integrating continuous testing throughout the cycle. In Kubernetes, continuous testing helps in identifying and addressing issues early, reducing the risk of significant problems at later stages. This approach ensures that each iteration is as stable and reliable as possible.

Flexibility and adaptability are essential characteristics of effective iterative cycles. The team should be prepared to modify their plans based on feedback received or unexpected challenges. This adaptability ensures that the cycle remains relevant and effective, even when faced with unforeseen circumstances.

Documentation plays a significant role in structuring these cycles. Maintaining detailed records of each iteration, including what was done, why it was done, and what the outcomes were, is invaluable for future reference and continuous learning.

Focusing on delivering tangible results at the end of each cycle is important. This focus helps in maintaining a sense of accomplishment and momentum, providing tangible benefits to the organization and end users.

Integrating user feedback into each cycle is crucial. Gathering and incorporating input from end users ensures that the development aligns with user needs and expectations, enhancing the overall effectiveness of the Kubernetes environment.

Ensuring a seamless transition between cycles is important for maintaining continuity and efficiency. This involves proper planning and preparation to ensure that the learnings and outputs from one cycle are effectively utilized in the next.

By structuring effective iterative cycles, Kubernetes teams can create a dynamic and responsive development environment. This approach not only enhances the quality and reliability of the Kubernetes implementation but also ensures that it evolves in line with user needs and organizational objectives.

## Case studies – iterative successes and failures

Examining case studies of iterative successes and failures provides valuable insights into the practical application of iterative development in Kubernetes environments. These case studies offer real-world examples of how this approach can lead to significant improvements, as well as cautionary tales of where it might go wrong.

One notable success story involves a company that adopted an iterative approach to refine its Kubernetes deployment. They started with a basic setup and, over several iterations, progressively incorporated more complex functionalities. This gradual process allowed them to manage risks effectively, as they could address issues as they arose without overwhelming their team or resources. The key to their success was their commitment to regular evaluation and adaptation, ensuring each iteration brought them closer to their desired state.

In contrast, a case of iterative failure demonstrates the importance of clear goal-setting and feedback integration. A different organization attempted to implement iterative changes to its Kubernetes infrastructure but lacked clear objectives for each cycle. Without these goals, their iterations became aimless, with changes implemented based on the latest trends rather than actual needs. Additionally, they failed to adequately incorporate feedback, leading to iterations that did not align with user expectations or resolve ongoing issues.

Another success involved a company focusing on automating its deployment process. By breaking down the automation process into smaller iterations, they managed to gradually transition from manual deployments to a fully automated pipeline. Each iteration allowed them to troubleshoot and refine their automation scripts, leading to a more reliable and efficient deployment process.

On the other hand, a failure in the iterative process can also stem from poor communication and collaboration. In one instance, a team working on a Kubernetes project operated in silos, with developers and operations teams working separately. This lack of collaboration led to iterations that often contradicted each other's efforts, causing delays and frustration. The lesson here highlights the need for cross-functional collaboration in successful iterative development.

A particularly instructive case study revolves around a company that successfully managed to scale its Kubernetes operations through iterative improvements. They initially faced performance issues at scale but addressed these problems through targeted iterations focused on optimizing their cluster configurations and resource allocations. Their success was largely due to a systematic approach to identifying and addressing specific bottlenecks in each cycle.

In terms of failures, another example involves a company that rushed its iterations without adequate testing. Eager to implement new features, they neglected thorough testing in each cycle, leading to stability and security issues. This case underscores the importance of balancing speed with quality assurance in the iterative process.

Reflecting on these case studies, the common factors in successful iterative developments include clear goal setting, regular feedback incorporation, effective communication and collaboration, and a balanced approach to risk management. Conversely, failures often result from a lack of direction, inadequate testing, poor communication, and ignoring user feedback. These real-world examples provide valuable lessons for organizations looking to adopt an iterative approach in their Kubernetes environments.

## Balancing speed and stability in iterations

Balancing speed and stability in iterations is crucial in Kubernetes management, ensuring that the swift pace of development doesn't undermine the system's reliability. This balance is achieved through several focused strategies.

Ensuring comprehensive testing at every stage of the iterative process is vital. It allows teams to swiftly detect and resolve issues, thus maintaining system stability. Automated testing proves particularly advantageous, as it efficiently conducts repetitive tests, enabling rapid development while upholding quality standards.

Setting realistic timelines is vital. Rapid development is important, but it should not compromise the thoroughness of planning, development, testing, and deployment. The pace should be brisk yet manageable, allowing careful execution at each phase.

Continuous monitoring and analysis of system performance are critical. This constant oversight helps to detect and rectify stability issues promptly, ensuring the system remains robust and responsive.

Version control and the ability to roll back to previous states are key in maintaining stability. If a new iteration introduces problems, teams can revert to a stable version, ensuring operational continuity.

Clear communication and collaboration among team members enhance the development pace. Effective communication leads to faster issue resolution and decision-making, crucial in a fast-paced environment.

Prioritizing updates and changes is another effective strategy. By focusing on the most impactful or urgent updates, teams can allocate their efforts more effectively, maintaining stability while advancing development.

Incorporating diverse perspectives and insights can guide each iteration. This approach involves understanding the implications of changes from various angles, ensuring that speed does not overshadow the need for a stable and reliable system.

Cultivating a team culture that values both rapid development and system stability is important. This culture ensures that all team members pursue speed with an understanding of its impact on stability.

By adopting these strategies, teams can maintain a delicate balance between rapid development and the stability of the Kubernetes environment. This balance is key to delivering timely, effective updates and maintaining a reliable, high-performing system.

## Tools and technologies supporting iterative practices

Tools and technologies play a pivotal role in supporting iterative practices. These tools facilitate efficient and effective development, testing, and deployment, allowing teams to embrace an iterative approach with confidence.

Container orchestration tools, led by Kubernetes, are foundational as they furnish the essential infrastructure for deploying and overseeing containerized applications on a large scale. Kubernetes, especially, furnishes features such as automated rollouts and rollbacks, self-healing, and scalability, all of which are indispensable for iterative development.

**Source code management** (**SCM**) tools such as Git are essential for version control. They enable teams to track changes, collaborate on code, and revert to previous versions if needed. This capability is vital for managing frequent updates and reversions that are often part of iterative development.

CI/CD tools are key enablers of iterative practices. Tools such as Jenkins, GitHub Actions, and GitLab CI automate the testing and deployment of code changes, facilitating rapid and frequent updates. They help in ensuring that each iteration is tested and deployed efficiently, reducing manual workload on teams.

Automated testing tools are indispensable in an iterative approach. Tools such as Selenium, JUnit, and others allow teams to create and run automated tests for their applications. These tests ensure that new code integrates seamlessly with existing code and meets quality standards.

Monitoring and logging tools such as Prometheus and the ELK (Elasticsearch, Logstash, Kibana) stack provide insights into application performance and system health. These tools are crucial for identifying issues early in the iterative process and for understanding the impact of changes on system performance.

Configuration management tools such as Ansible help in automating the configuration of servers and other infrastructure. This automation is essential for maintaining consistency and reliability, especially when frequent changes are made during iterative development.

Containerization tools such as Docker play a significant role. They allow applications to be packaged with their dependencies, ensuring consistency across different environments. This consistency is key in iterative development, where applications need to be deployed frequently under varying conditions.

Cloud-based development environments and services offer flexibility and scalability, which are beneficial for iterative practices. Cloud platforms such as Amazon Web Services (AWS), Azure, and Google Cloud provide a range of services that support Kubernetes and containerization, making it easier for teams to deploy and manage their applications.

Artifact repositories such as JFrog Artifactory and Nexus are important for storing and managing build artifacts. They provide a centralized location for artifacts, making it easier to manage the outputs of the development process across different iterations.

Collaboration and project management tools such as Slack, Jira, and Trello facilitate effective team communication and organization. These tools help in tracking progress, assigning tasks, and ensuring that everyone is aligned with the project's goals and timelines.

By leveraging these tools and technologies, teams working with Kubernetes can adopt and enhance their iterative practices. This adoption leads to more efficient development cycles, higher-quality outputs, and ultimately, a more robust and scalable Kubernetes environment.

## Iterative planning and roadmapping

Iterative planning and roadmapping involve breaking down the project into smaller, manageable segments, allowing for flexibility and adaptability as the project evolves.

The process starts by outlining the overarching vision and long-term objectives of the Kubernetes project. This initial phase establishes the direction and purpose, which then directs the following planning stages. It's crucial to comprehend precisely what the project aims to accomplish and how it fits into broader organizational goals.

Next, the project is broken down into smaller iterations or phases. Each of these iterations should have specific, achievable objectives. This breakdown makes the project more manageable and allows for frequent reassessments and adjustments. It's crucial that these objectives are clear and measurable, providing a concrete basis for evaluating progress.

Setting realistic timelines for each iteration is essential. These timelines should account for the complexity of tasks, dependencies between tasks, and potential risks. A well-thought-out timeline helps maintain a steady pace of development and ensures that the team has enough time to complete each task to the required standard.

Involving the entire team in the planning process is beneficial. This collaborative approach ensures that different perspectives are considered, leading to a more comprehensive plan. It also ensures that all team members are on the same page and understand their roles and responsibilities in each iteration.

Regularly reviewing and updating the roadmap is a critical aspect of iterative planning. As the project progresses, new information, changing requirements, or unforeseen challenges may arise. Regular reviews allow the team to adapt their plan to these changes, ensuring that the project remains on track and relevant.

Prioritizing tasks within each iteration is another important step. Not all tasks have the same level of importance or urgency. By prioritizing tasks, teams can focus their efforts on what's most critical, ensuring efficient use of resources and time.

Incorporating feedback from previous iterations is a key part of the planning process. Lessons learned from earlier phases should inform future planning, helping to avoid past mistakes and capitalize on successful strategies.

Risk assessment and mitigation should be integrated into the planning process. Identifying potential risks early on and planning for them can save a lot of time and resources later. This approach ensures that the project remains resilient and adaptable.

Communication of the plan and roadmap to all stakeholders is crucial. Keeping everyone informed not only fosters transparency but also ensures alignment and support from across the organization.

Being flexible and open to change is essential. Iterative planning is not about sticking rigidly to a plan but about adapting to new information and circumstances. This flexibility is key to managing a successful Kubernetes project.

Iterative planning and roadmapping, when executed effectively, lead to a more controlled and adaptable project management process. This approach not only helps in achieving the immediate goals of each iteration but also ensures that the overall project remains aligned with its long-term objectives.

## Feedback integration in iterative processes

Integrating feedback effectively into iterative processes ensures that each iteration not only meets technical requirements but also aligns with user expectations and business objectives. This integration process involves several key steps and strategies.

Establishing clear channels for gathering feedback is essential, involving various methods such as user surveys, direct customer interviews, or feedback from internal teams. Additionally, performance metrics and logs from the Kubernetes system itself can provide valuable insights into how changes are affecting the system's performance and stability.

Once feedback is collected, it's important to systematically analyze and prioritize it. Not all feedback will have the same level of urgency or impact. Teams need to assess feedback based on its potential to improve the system and align it with the project's overall goals. This prioritization helps in focusing efforts on the most impactful changes.

Incorporating feedback into the planning phase of each iteration is a critical step. This planning should involve revising goals and objectives based on the feedback received. It may also require redefining the scope of work for the upcoming iteration to address key issues or incorporate new requirements.

Effective communication of how feedback is being used is also vital. Stakeholders and team members appreciate understanding how their input is making a difference. This transparency can increase engagement and trust in the process, leading to more constructive and actionable feedback in future cycles.

Another important aspect is the iterative testing of changes based on feedback. As new changes are implemented, they should be tested not only for technical performance but also for how well they address feedback received. This testing can be part of the CI/CD pipeline.

Iterative reviews and retrospectives offer an opportunity to reflect on how feedback was integrated and what the outcomes were. These reviews can provide insights into the effectiveness of the feedback integration process and highlight areas for improvement.

Adapting the feedback integration process over time is also important. As the project evolves, the type of feedback and methods of collecting and integrating it may need to change. Being flexible and open to adjusting the process ensures that feedback integration remains effective throughout the life cycle of the Kubernetes project.

Maintaining a customer-centric approach throughout the iterative process ensures that feedback integration remains a top priority. Keeping the end user in mind at every stage of the process helps in making decisions that enhance the overall value and usability of the Kubernetes environment.

By effectively integrating feedback into iterative processes, Kubernetes teams can ensure that their projects are not only technically sound but also closely aligned with user needs and business goals. This approach leads to more successful outcomes and a Kubernetes environment that continually evolves to meet changing demands.

## Iterative risk management and mitigation strategies

Employing iterative risk management and mitigation strategies in Kubernetes is essential to continuously identify, assess, and address risks throughout the development process.

Ensuring early identification of potential risks is vital for effective risk management. It requires a thorough analysis of every iteration to pinpoint possible issues, such as code vulnerabilities and infrastructure inadequacies. Proactively identifying these risks helps prevent them from developing into major problems.

Once risks are identified, they need to be assessed in terms of their potential impact and likelihood. This assessment helps in prioritizing risks, focusing attention on those that could have the most significant effect on the project. High-impact risks require more immediate and detailed attention.

Developing a mitigation plan for each identified risk is crucial. These plans should outline the steps to either reduce the likelihood of the risk occurring or minimize its impact if it does occur. For instance, backup strategies might be implemented to mitigate the risk of data loss.

Implementing these mitigation strategies as part of the iterative process is essential. By incorporating risk management into regular development cycles, teams can ensure that they are continually addressing potential issues. This ongoing attention to risk helps in maintaining the stability and security of the Kubernetes environment.

Regularly revisiting and updating risk assessments is important. As the project evolves, new risks may emerge, and existing risks may change in nature. Regular reviews ensure that the risk management strategy remains relevant and effective.

Documentation plays a key role in this process. Keeping detailed records of identified risks, assessments, and mitigation actions provides a clear history that can be referred to in future iterations. This documentation is invaluable for understanding past challenges and how they were addressed.

Communication about risks and mitigation strategies is vital. All team members should be aware of potential risks and steps being taken to manage them. This transparency ensures that everyone is prepared to respond appropriately if an issue arises.

In addition to proactive risk management, having a reactive plan in place is necessary. Despite the best efforts, some risks will materialize. A reactive plan outlines the steps to take when this happens, helping to minimize disruption and quickly restore normal operations.

Focusing on training and awareness is also a key strategy. Educating team members about common risks in Kubernetes environments and how to avoid or mitigate them can significantly reduce the likelihood of issues occurring.

Leveraging automation where possible can enhance risk management. Automated tools can monitor the system for signs of potential issues, perform regular security scans, and even implement certain mitigation strategies automatically.

By incorporating these iterative risk management and mitigation strategies into their workflows, Kubernetes teams can create a more secure and stable environment. This approach not only addresses immediate risks but also contributes to the long-term health and success of the project.

## Adapting to the evolving Kubernetes ecosystem

This section addresses tracking and responding to Kubernetes ecosystem changes, embracing new features and updates, the role of community and collaboration in adaptation, adapting deployment strategies for new challenges, continuous security practices, managing dependencies, predicting future trends in Kubernetes development, and building a resilient mindset for technological evolution.

# Tracking and responding to Kubernetes ecosystem changes

Effectively managing Kubernetes environments requires strategies for tracking and responding to changes within the Kubernetes ecosystem. This involves implementing various practices and approaches to maintain the system's currency, security, and efficiency.

Regular engagement with the Kubernetes community and industry sources is essential. This includes participating in forums, attending conferences, and subscribing to relevant newsletters. Such engagement offers insights into emerging trends, best practices, and upcoming changes that could impact Kubernetes environments.

Staying updated with official Kubernetes releases and updates is no longer an option. Teams should monitor release schedules and notes provided by the Kubernetes project. This information is critical for understanding new features, bug fixes, security patches, and any deprecated functionalities that might require attention.

Implementing a system for monitoring technological advancements related to Kubernetes can significantly aid in timely responses to changes. Tools that track specific keywords or topics related to Kubernetes on various platforms can provide early alerts about new developments.

Regular audits and reviews of the current Kubernetes setup are important. These reviews help identify areas that may require updates or improvements in light of new changes in the ecosystem. They ensure that the system remains optimized and aligned with the latest standards.

Training and development for team members are key to keeping pace with the ecosystem's changes. Encouraging continuous learning and providing resources for training helps in building a knowledgeable team capable of adapting to new technologies and practices.

Creating a strategic plan for integrating new changes into the existing Kubernetes environment is beneficial. This plan should include assessing the impact of changes, testing new features in a controlled environment, and developing a rollout strategy that minimizes disruption.

Developing close relationships with vendors and partners who specialize in Kubernetes can provide additional support. These relationships can offer access to specialized knowledge and insights, helping to navigate complex changes more effectively.

Incorporating user feedback in response to changes in the Kubernetes ecosystem is also important. User feedback can provide practical insights into how changes affect the usability and performance of the system.

Keeping a balance between embracing new advancements and maintaining system stability is important. While it's necessary to use new features and improvements, it's equally vital to make sure these changes don't harm the system's integrity.

By employing these strategies, teams can effectively track and respond to changes in the Kubernetes ecosystem, ensuring their environments remain up to date, secure, and optimized for performance.

## Embracing new Kubernetes features and updates

Adopting new features and updates in Kubernetes is an important process that keeps the system efficient, secure, and up to date with technological advancements. This involves a series of steps and considerations to effectively integrate new developments.

Understanding the specifics of each new feature or update is essential. It requires reading through the release notes and documentation provided by Kubernetes. By grasping the benefits, potential limitations, and use cases of new features, teams can make informed decisions about which updates to adopt.

Evaluating the compatibility of new features with the existing Kubernetes environment is crucial. This evaluation should consider how new updates will interact with the current setup, including applications, integrations, and custom configurations. Compatibility checks help prevent conflicts and ensure seamless integration.

Testing new features in a controlled environment before full-scale implementation is key. This can be done in a staging or development environment that mimics the production setup. Testing helps identify any issues or adjustments needed and assesses the impact of updates on overall system performance.

Planning a phased rollout of new features is often a wise approach. Instead of implementing updates across the entire system at once, gradually introducing changes allows for closer monitoring and reduces the risk of widespread issues. This phased approach also provides the flexibility to roll back or adjust the plan if needed.

Training and knowledge sharing among team members about new features and updates are vital. Organizing training sessions, workshops, or knowledge-sharing meetings helps ensure that all team members are up to date and can effectively work with new Kubernetes features.

Monitoring and analyzing the impact of new features post-implementation is important. After integrating new updates, continuous monitoring helps track their performance and impact. This monitoring provides valuable feedback and informs future decisions on adopting and utilizing Kubernetes features.

Engaging with the Kubernetes community can provide additional insights and support. Community forums, user groups, and online discussions can be excellent resources for tips, best practices, and troubleshooting advice related to new features.

Maintaining documentation of changes and updates within the Kubernetes environment is also beneficial. Keeping detailed records of what changes were made, why, and their outcomes helps in maintaining a clear history of the system's evolution and can be a valuable resource for troubleshooting and future planning.

Staying flexible and adaptable to change is paramount. The Kubernetes ecosystem is constantly evolving, and new features or updates might require shifts in strategy or approach. Being open to change and adaptable in planning and execution ensures that teams can effectively leverage new developments in Kubernetes.

## The role of community and collaboration in adaptation

The role of community and collaboration in adapting to the evolving Kubernetes ecosystem is incredibly valuable. Engaging with the broader Kubernetes community and promoting collaborative efforts within and across teams can significantly improve the ability to effectively navigate and utilize changes in the ecosystem.

Engagement in forums, mailing lists, **special interest groups** (**SIGs**), and attending Kubernetes-focused events such as *KubeCon* is crucial. Such involvement provides access to a wealth of knowledge, experiences, and insights from a diverse range of users and contributors. It enables teams to remain updated on best practices, emerging trends, and common challenges faced by others in the field.

Collaboration with other teams and organizations is also crucial. Sharing experiences and solutions with peers can provide new perspectives and innovative approaches to common problems. This collaboration can take various forms, such as joint workshops, co-development initiatives, or simply regular knowledge exchange sessions.

Internal collaboration within organizations plays a significant role in adapting to changes in Kubernetes. Encouraging open communication and cross-functional teamwork ensures that different perspectives and expertise are brought together. This collaborative environment is essential for effectively assessing, planning, and implementing changes in the Kubernetes ecosystem.

Leveraging online resources and platforms dedicated to Kubernetes can enhance collaborative efforts. Websites, forums, and social media groups focused on Kubernetes serve as platforms for discussion, problem-solving, and knowledge sharing. These resources can be particularly valuable for keeping up with rapid developments and for seeking advice on specific challenges.

Contributing back to the Kubernetes community is another important element. By sharing experiences, code contributions, or even documentation improvements, teams can give back to the community that supports them. This contribution not only enriches the community but also helps in building a positive reputation and network within the ecosystem.

Fostering a culture of continuous learning within teams is essential for keeping pace with the evolving Kubernetes landscape. Encouraging team members to engage in ongoing education, whether through formal training, self-study, or community events, ensures that the collective skill set remains current and diverse.

Creating internal forums or groups focused on Kubernetes allows team members to share insights, ask questions, and discuss challenges related to Kubernetes. These internal communities can act as a support network and a hub for collective problem-solving.

Partnering with Kubernetes experts or consultants can provide additional support and guidance. These experts can offer specialized knowledge and experience, helping teams navigate complex changes or adopt new practices more effectively.

By engaging with the wider community, fostering internal collaboration, and continuously learning, teams can effectively navigate changes, share knowledge, and collectively enhance their Kubernetes practices. This collaborative approach not only benefits individual teams and organizations but also contributes to the strength and vitality of the broader Kubernetes community.

## Adapting deployment strategies for new challenges

Adjusting deployment strategies to meet new challenges in Kubernetes environments is crucial for maintaining efficiency, security, and performance. This adaptation involves assessing and adjusting existing deployment processes based on changing requirements, technological advancements, and emerging best practices.

Understanding the nature of new challenges is the initial step. This could involve changes in application requirements, updates in Kubernetes itself, shifts in user expectations, or emerging security threats. A clear understanding of these challenges aids in formulating effective adaptation strategies.

Revising containerization practices may be necessary. As applications evolve, their dependencies and configurations might change, necessitating updates in how containers are built and managed. This revision might include optimizing Dockerfiles, updating base images, or adopting new container technologies.

Modifying CI/CD pipelines to accommodate new requirements is often crucial. As deployment processes evolve, CI/CD workflows may need adjustments. This could involve integrating new testing tools, automating additional steps, or reconfiguring pipelines for increased efficiency.

Scaling strategies might need to be revisited. Kubernetes offers various scaling options, including horizontal pod autoscaling and cluster autoscaling. Adapting these strategies to respond to changing traffic patterns or workload characteristics ensures that resources are utilized optimally.

Enhancing security measures in deployment processes is vital, especially in response to new vulnerabilities or compliance requirements. This might involve implementing more robust authentication and authorization practices, encrypting data in transit and at rest, or integrating advanced security scanning tools.

Optimizing resource management can improve deployment efficiency. This involves fine-tuning resource requests and limits for pods, leveraging more efficient storage solutions, or adopting cost-effective cloud services.

Incorporating advanced deployment techniques, such as canary releases, blue-green deployments, or feature flags, can mitigate risk. These techniques allow for gradual rollouts and easier rollback, reducing the impact of potential issues in new deployments.

Monitoring and observability should be enhanced to provide deeper insights into the deployment process and application performance. Advanced monitoring tools can help identify issues early and provide data-driven insights for further optimization.

Staying informed about the latest Kubernetes features and community best practices is also important. Regularly updating knowledge and skills ensures that the deployment strategies remain current and effective.

Regularly reviewing and updating documentation ensures that the entire team has access to the latest information on deployment processes and strategies. Well-maintained documentation is crucial for consistency and efficiency, especially in fast-evolving environments.

## Continuous security practices in a changing ecosystem

In a constantly changing Kubernetes ecosystem, keeping up with security practices is vital to defend against new threats and vulnerabilities. This continuous security effort includes various key strategies to guarantee the security and resilience of the Kubernetes environment against emerging challenges.

These policies should cover all aspects of the Kubernetes environment, from access control and network policies to resource limitations and pod security. Regularly reviewing and updating these policies in response to new threats or best practices in the ecosystem is crucial.

Automating security processes wherever possible enhances efficiency and consistency. Tools that automate security scanning, patch management, and compliance checks can significantly reduce the risk of human error and ensure that security measures are applied uniformly across the entire environment.

Continuous monitoring and logging of security events are vital for early detection of potential threats. Monitoring solutions should be configured to track unusual activities, such as unauthorized access attempts or unexpected changes in resource usage. This continuous vigilance allows for a swift response to potential security incidents.

Regular vulnerability assessments and penetration testing are important to identify and address weaknesses in the Kubernetes environment. These assessments should be conducted periodically and whenever significant changes are made to the system, ensuring that new updates or configurations do not introduce vulnerabilities.

Staying informed about the latest security threats and trends in the Kubernetes ecosystem is essential. Subscribing to security bulletins, participating in Kubernetes security forums, and attending relevant conferences can provide valuable insights into emerging threats and recommended protective measures.

Educating and training team members on security best practices is also crucial. Regular training sessions, workshops, and security drills can help ensure that all team members are aware of the latest security risks and know how to respond effectively.

Developing a robust **incident response plan** (**IRP**) is necessary for dealing with security breaches or vulnerabilities effectively. This plan should outline clear procedures for responding to different types of security incidents, including who to contact, how to isolate affected systems, and how to communicate with stakeholders.

Integrating security considerations into the development and deployment processes helps prevent vulnerabilities from being introduced into the environment. This includes conducting security reviews of code and configurations, as well as integrating security testing into the CI/CD pipeline.

Collaborating with external security experts or vendors can provide additional support and expertise. These partners can offer specialized knowledge, tools, and services to enhance the security of the Kubernetes environment.

Finally, fostering a culture of security within the organization is important. Encouraging a mindset where security is everyone's responsibility and promoting open communication about security concerns can lead to a more proactive and vigilant approach to security practices.

By implementing these continuous security practices, organizations can ensure that their Kubernetes environments remain secure and resilient in the face of a constantly changing ecosystem. This proactive and comprehensive approach to security is essential for protecting against current threats and preparing for future challenges.

## Managing dependencies in a dynamic environment

Effectively managing dependencies in a dynamic Kubernetes environment is vital for maintaining system stability and efficiency. Dependencies can greatly affect application performance and reliability, so implementing effective management strategies is crucial to address them in a constantly evolving ecosystem.

One key strategy is implementing an automated dependency management system. Tools such as Helm for Kubernetes can manage complex dependencies, automate deployment processes, and ensure that the right versions of applications and their dependencies are used. Automation reduces the risk of human error and simplifies the management process.

Regularly auditing and updating dependencies is important. This involves keeping track of dependencies each application uses and regularly checking for updates or patches. Staying current with the latest versions can prevent security vulnerabilities and ensure compatibility with the Kubernetes environment.

Establishing clear policies for dependency management is beneficial. These policies should define how to add new dependencies, the process for updating them, and the criteria for selecting third-party libraries or services. Clear guidelines help maintain consistency and reduce the risk of introducing problematic dependencies.

Using containerization effectively helps isolate dependencies. By packaging applications with their dependencies in containers, conflicts between different applications or different parts of the same application can be minimized. This isolation simplifies dependency management and enhances the stability of the environment.

Implementing version control rigorously is crucial. Proper version control practices ensure that changes to dependencies are tracked, making it easier to revert to previous versions if an update causes issues. This practice is essential for maintaining a stable and functional environment.

Testing is a critical component of managing dependencies. Automated testing should be used to validate that updates to dependencies do not break the application. Integration tests, in particular, can ensure that the application works as expected with the updated dependencies.

Monitoring the performance impact of dependencies is also necessary. Sometimes, updates to dependencies can affect application performance. Continuous monitoring can help quickly identify and address performance issues that arise due to dependency changes.

Documenting dependencies and their impact is vital for future reference and for new team members. Documentation should include information about why a particular dependency is used, how it interacts with the application, and any special considerations for its maintenance.

Collaborating with the wider community can provide insights into how others manage dependencies. Engaging in forums, attending meetups, or participating in open source projects can offer valuable tips and best practices.

Planning for dependency deprecation is important as dependencies can become deprecated or unsupported over time. Having a plan for replacing or updating these dependencies ensures that the application remains secure, stable, and up to date.

Through prioritizing these strategies, teams can proficiently handle dependencies in a dynamic Kubernetes environment, minimizing risks linked with dependency issues and guaranteeing the seamless operation of applications.

## Predicting future trends in Kubernetes development

Anticipating future trends in Kubernetes development requires analyzing current patterns, technological advancements, and the evolving needs of organizations. By staying ahead of these trends, teams can better prepare for changes and opportunities that lie ahead in the Kubernetes ecosystem.

One key trend is the increasing focus on simplicity and user-friendliness. As Kubernetes becomes more mainstream, there's a growing emphasis on making it more accessible to a broader range of users, including those who may not have deep technical expertise in container orchestration. This could mean more intuitive interfaces, simplified management tools, and enhanced automation to reduce the complexity of deploying and managing Kubernetes.

The integration of AI and **machine learning** (**ML**) into Kubernetes is likely to continue gaining momentum. These technologies can be used to enhance various aspects of Kubernetes, such as optimizing resource allocation, improving security through predictive analytics, and automating routine tasks. This integration will make Kubernetes smarter and more efficient.

Edge computing is expected to become more prominent in Kubernetes development. As the volume of data generated at the edge of networks continues to grow, Kubernetes will likely evolve to better support edge computing scenarios. This includes managing deployments across a distributed infrastructure and ensuring seamless operation across cloud and edge environments.

Security will remain a top priority, with ongoing efforts to make Kubernetes environments more secure. This might involve developing more robust built-in security features, enhanced encryption techniques, and tighter integrations with existing security tools and frameworks.

The trend toward hybrid and multi-cloud deployments will likely continue. Kubernetes is well positioned to be the orchestrator of choice for these environments, thanks to its ability to run consistently across different cloud providers. Future developments in Kubernetes may focus on improving its capabilities in managing resources and applications across various clouds seamlessly.

Serverless computing is another area where Kubernetes could see significant developments. As the demand for serverless options grows, Kubernetes might evolve to offer better support for serverless architectures, enabling organizations to run applications without managing the underlying infrastructure.

Sustainability and eco-friendly computing might emerge as a new focus area. This could involve optimizing Kubernetes to be more energy-efficient, reducing its carbon footprint, and supporting green computing initiatives.

The growth of service mesh technology, which enhances the capabilities of Kubernetes by managing complex service-to-service communications, is expected to continue. Future Kubernetes releases may offer deeper integrations with service mesh technologies, providing out-of-the-box solutions for advanced networking, security, and observability features.

Community-driven innovation will continue to shape Kubernetes. The open source nature of Kubernetes means that its development is influenced by a wide range of contributors, from individual developers to large corporations. This collaborative approach will drive diverse innovations and ensure that Kubernetes remains at the forefront of container orchestration technology.

## Building a resilient mindset for technological evolution

Developing a resilient mindset for technological evolution, particularly in the context of Kubernetes and its rapidly changing landscape, is crucial for teams and organizations to adapt and thrive. This mindset encompasses several key attitudes and approaches that aid individuals and teams in navigating technological changes effectively.

Encouraging adaptability is another important aspect. Teams must be prepared to adjust their strategies and plans as new technologies and updates emerge. This adaptability ensures that they can quickly take advantage of new opportunities and mitigate potential challenges brought about by changes in the ecosystem.

Promoting a culture of experimentation and innovation within teams is also vital. Encouraging team members to experiment with new tools, technologies, and processes can lead to valuable insights and breakthroughs. This culture of innovation helps teams find novel solutions to emerging challenges and stay ahead in a competitive landscape.

Developing strong problem-solving skills is crucial for resilience. As technologies evolve, new types of challenges and issues arise. Teams equipped with robust problem-solving skills can navigate these challenges more effectively, turning potential obstacles into opportunities for growth and improvement.

Emphasizing the importance of collaboration and knowledge sharing helps build a supportive environment. In the face of technological evolution, sharing experiences, insights, and learnings within and between teams can significantly enhance collective understanding and capabilities.

Maintaining a positive attitude toward change is key. Viewing technological evolution as an opportunity rather than a threat can transform how teams approach new developments. This positive perspective fosters a more open and proactive approach to learning and adaptation.

Staying connected with the broader tech community, including Kubernetes user groups, forums, and conferences, provides a broader perspective. These connections offer insights into how others are adapting to changes, providing inspiration and practical ideas for one's own context.

Balancing a focus on both short-term needs and long-term visions is important. While it's necessary to address immediate challenges and opportunities presented by new technologies, keeping an eye on the future ensures that decisions and strategies are aligned with long-term goals and trends.

Practicing resilience in the face of failures and setbacks is essential. In a rapidly evolving technological landscape, not every initiative or project will be successful. Learning from these experiences and using them as stepping stones for future efforts is a hallmark of a resilient mindset.

## Summary

This chapter provided an in-depth exploration of continuous improvement in Kubernetes environments, emphasizing its critical role in adapting to the rapidly changing technology landscape. Foundational concepts of continuous improvement were examined, along with the significance of incorporating feedback into iterative processes. Traditional models were compared with modern continuous improvement methodologies. The importance of measuring the success of continuous improvement initiatives and the psychological aspects of fostering a growth mindset were discussed. Practical aspects such as continuous learning, aligning continuous improvement with DevOps practices, and effective risk management in iterative contexts were thoroughly explored. Also, guidance was offered on adapting to evolving changes in the Kubernetes ecosystem, including embracing new features, updates, and the essential role of community and collaboration.

In the next chapter, we will explore proactive assessment and prevention in Kubernetes environments, examining the significance of developing a proactive mindset, anticipating potential pitfalls, and implementing preventive measures to uphold system stability and security.

# 8

# Proactive Assessment and Prevention

This chapter highlights proactive assessment and prevention in Kubernetes, emphasizing the importance of a forward-looking approach in cloud computing. It discusses developing a proactive mindset, early detection for system health, and strategic planning influenced by Kubernetes versioning. This chapter also touches on sustainability in design, adaptive leadership, and preparing for challenges with risk assessments and scenario planning. Key focuses include implementing preventive measures such as documentation, disaster simulations, financial governance, and compliance protocols. Finally, it offers insights into strategic growth management in Kubernetes, ensuring scalability and resilience through proactive steps.

We'll cover the following topics in this chapter:

- Developing a proactive Kubernetes mindset
- Assessing and anticipating potential pitfalls
- Implementing preventative measures

## Developing a proactive Kubernetes mindset

This section addresses developing a proactive mindset in Kubernetes management, focusing on prevention over correction, grasping ecosystem trends, and the critical role of early detection. It also explores the psychological foundations of proactive strategies, the impact of versioning on planning, the significance of sustainability in design, and the importance of adaptive leadership.

## The importance of proactivity in Kubernetes management

Embracing a proactive approach in managing Kubernetes environments plays a critical role in navigating the complexities and dynamic challenges presented by cloud-native technologies. This strategy is centered on the anticipation of issues before they manifest into significant problems, underlining the need for a deep understanding of your Kubernetes setup. By leaning into proactive measures, teams are empowered to identify vulnerabilities, optimize the use of resources, and significantly improve the performance and security of their systems.

A foundational element of this proactive stance is the commitment to regular, comprehensive monitoring. This involves keeping a vigilant eye on the system to catch anomalies as early as possible, enabling quick responses to mitigate potential risks. However, effective monitoring transcends the technical aspects, encompassing a review of operational practices to ensure they align with the goal of maintaining optimal system performance.

Predictive analytics stands as another cornerstone of proactive Kubernetes management. Through careful analysis of data trends and usage patterns within Kubernetes clusters, teams can anticipate future needs and challenges. This foresight proves invaluable in managing resources efficiently, ensuring that applications maintain high responsiveness and availability without unnecessary expenditure on resources.

Investing in the training and continuous education of the team is also paramount. A team that is up-to-date with the latest Kubernetes trends, challenges, and best practices is better equipped to make informed decisions. Such a team is proactive by nature and capable of identifying potential issues well before they become critical, fostering an environment where continuous improvement is the norm.

Moreover, a proactive mindset toward Kubernetes management requires a dedication to the principle of continuous refinement. Organizations must regularly assess and enhance their strategies, processes, and configurations to stay ahead of the curve. This involves not just reacting to the current landscape but also preparing for future developments in Kubernetes and cloud-native technologies.

By integrating diligent monitoring with predictive analytics, prioritizing team education, and committing to ongoing strategy and process improvement, organizations can develop and maintain a proactive mindset in Kubernetes management. This approach not only secures and optimizes Kubernetes environments but also cultivates a culture of proactive problem-solving and innovation. Such a culture is invaluable in the rapidly evolving domain of cloud computing, where the ability to anticipate and adapt to changes can significantly impact the success and resilience of cloud-native applications.

## Building a culture of prevention over correction

Shifting from a reactive to a preventive strategy represents a pivotal change in the management of Kubernetes environments. This evolution toward a culture that prioritizes prevention over correction highlights the crucial role of identifying and addressing potential issues before they escalate into significant challenges. It cultivates an organizational mindset that values foresight and meticulous planning, aiming to circumvent potential problems rather than expending resources on resolving issues after they have occurred.

Initiating this cultural shift begins with leadership. Leaders must set an example and establish clear expectations for proactive behavior within their teams. They play a vital role in advocating for the preventive approach, illustrating its merits through their actions and strategic choices. By embedding preventive measures into both strategic planning and day-to-day operations, leaders can disseminate the importance of this proactive approach throughout the organization, ensuring it permeates every level.

At the core of fostering this culture are education and training. Regular, targeted training sessions focused on Kubernetes best practices, the identification of potential risks, and strategies for their anticipation and mitigation are indispensable. These educational initiatives should extend beyond mere technical training to also encourage a mindset oriented toward foresight and the prevention of issues before they arise.

Another key element in this strategic shift is the establishment of comprehensive review and planning processes. These processes must include systematic reviews of the Kubernetes environment to uncover any potential vulnerabilities or areas of inefficiency. Subsequent planning efforts should then concentrate on addressing these findings proactively, whether that involves modifications to configurations, the introduction of new tools, or adjustments in operational practices to avert potential issues.

Encouraging open communication and collaboration across teams is also crucial in strengthening this culture of prevention. When team members feel empowered to share their insights and concerns regarding potential issues, the organization can take collective, preemptive action to address them. This emphasis on collaborative problem-solving fosters a more cohesive and unified preventive strategy across the organization.

Recognizing and rewarding proactive behavior is another essential component. Acknowledging teams or individuals who proactively identify and address potential issues before they escalate not only highlights the value of such behaviors within the organization but also serves as motivation for others to adopt similar proactive practices. This recognition can be manifested in various forms, ranging from formal awards to informal commendations during team meetings.

By dedicating efforts to cultivating a culture that emphasizes prevention over correction, organizations can significantly mitigate the frequency and severity of issues within their Kubernetes environments. Adopting this proactive approach not only leads to smoother, more efficient operations but also fosters an organizational ethos that is forward-thinking and adept at navigating future challenges, thereby ensuring a more resilient and proactive organizational framework.

## Understanding Kubernetes ecosystem trends

Understanding the evolving landscape of Kubernetes is crucial for organizations committed to maintaining a proactive stance in their operational strategies. This deep understanding allows teams to anticipate changes in technology, adapt their practices, and foresee potential challenges, enabling them to adjust their plans and tactics accordingly. As Kubernetes evolves, its ecosystem expands, introducing new tools, features, and methodologies that significantly impact how deployments are managed and optimized.

Staying informed about developments directly related to Kubernetes is essential. This includes staying updated on changes to the core platform, new releases, and the removal of existing features. Being prepared for these changes involves testing new versions in controlled environments, understanding the benefits of new features, and planning migrations away from deprecated functionalities.

Exploring the growing array of Kubernetes-native tools and services is another vital aspect of staying informed. The ecosystem offers various solutions designed to enhance observability, security, and networking. By evaluating and integrating these tools into their workflows, teams can leverage advancements to improve the efficiency, resilience, and overall performance of their Kubernetes environments.

Active involvement in the community and keeping track of industry trends are also crucial. The Kubernetes community, which is bustling with users, developers, and vendors, provides valuable insights, best practices, and innovative applications of the platform. Participating in community forums, attending relevant conferences, and contributing to open source projects offer opportunities to stay updated on the ecosystem's dynamics, emerging trends, challenges, and innovative solutions.

By actively participating in Kubernetes SIGs, organizations not only keep up with trends but also contribute to the evolution of Kubernetes. Organizations should start by identifying which SIGs align with their interests or areas of expertise. This could be a SIG related to their business operations, such as SIG-Apps for application development or SIG-Security for enhancing security measures. As organizations get more involved, there are opportunities to take on leadership roles within a SIG. This could mean becoming a SIG chair or tech lead, positions that have significant influence over the direction of SIG projects.

It's essential to understand Kubernetes within the broader context of cloud computing, DevOps practices, and software development methodologies. Trends such as the increasing adoption of serverless architectures, the shift toward GitOps for infrastructure management, and the heightened focus on security in the software supply chain significantly influence Kubernetes management and utilization.

Learning from adoption patterns and case studies from other organizations provides additional insights. Drawing from the experiences of others, including their successes and challenges, offers valuable perspectives. This collective knowledge can inform strategic decision-making, help you navigate common pitfalls, and inspire innovative approaches to effectively leveraging Kubernetes.

A comprehensive approach is necessary to fully grasp the trends within the Kubernetes ecosystem. This approach should encompass vigilantly monitoring the platform's evolution, actively engaging with the community, exploring new tools, and learning from the broader industry landscape. Armed with this knowledge, organizations can not only refine their current Kubernetes deployments but also strategically position themselves for future advancements, ensuring ongoing technological leadership and innovation.

## Early detection – the key to Kubernetes health

Early detection is crucial for ensuring the well-being and reliability of Kubernetes environments. By spotting potential issues before they escalate, teams can maintain the strength and performance of their deployments. This proactive approach forms the bedrock of effective Kubernetes management, especially in navigating the intricate and dynamic ecosystem where underlying issues may remain hidden until they reach critical levels.

One essential strategy for achieving early detection is implementing comprehensive monitoring and alert systems. These systems offer real-time insights into the operational status of Kubernetes clusters, covering aspects such as resource usage, performance metrics, and system health. By setting up alerts for abnormal behavior or threshold breaches, teams can promptly address potential issues upon detection.

Another vital aspect involves utilizing log aggregation and analysis tools. These tools gather logs from various Kubernetes components, enabling teams to efficiently analyze vast amounts of data. By scrutinizing logs, teams can uncover patterns or irregularities that indicate emerging issues, allowing for proactive intervention.

Regular vulnerability scans and security audits are also crucial for the early detection of security-related issues. Due to Kubernetes' open nature and complex dependencies, vulnerabilities can arise from multiple sources. Proactive scanning and auditing help identify these vulnerabilities early, reducing the risk of potential exploits.

Integrating performance testing and benchmarking into the **continuous integration and delivery (CI/CD)** pipeline can also aid in the early detection of performance regressions or bottlenecks. By automating these tests, teams can evaluate the impact of code base or infrastructure changes on system performance before deployment to production.

Engagement with the Kubernetes community and staying informed about known issues and patches are equally vital. Often, issues that are encountered by one organization have already been addressed by others in the community. Leveraging this collective knowledge accelerates issue identification and resolution within your environment.

However, early detection isn't just about tools and processes; it's also about nurturing a culture of vigilance and continuous improvement. Encouraging team members to remain alert to potential issues and prioritize their swift resolution underscores the importance of maintaining system health and stability. Through a blend of technological solutions and organizational commitment, early detection becomes a cornerstone in ensuring the long-term vitality of Kubernetes deployments.

## The psychological aspect of proactive Kubernetes management

The psychological dimension of proactive Kubernetes management is frequently underestimated, yet it holds a pivotal role in ensuring the seamless operation of cloud-native environments. Beyond the technical intricacies, it encompasses the intricate interplay of team dynamics, individual mindsets, and organizational culture, all of which significantly influence the ability to anticipate and address potential issues before they escalate.

Taking a proactive stance toward Kubernetes management necessitates a fundamental shift in mindset from reactive problem-solving to proactive anticipation and prevention. This transition demands a level of foresight and strategic planning that can be challenging to cultivate within the fast-paced and dynamic operational landscapes characteristic of cloud-native environments. It calls for the cultivation of a culture of perpetual vigilance, where teams are willing to invest time and resources into preventive actions, recognizing their long-term benefits for system stability and performance.

At the heart of fostering this proactive mindset lies leadership. Leaders play a pivotal role in setting the tone and example for the entire organization. By prioritizing strategic planning, investing in training initiatives, and championing the adoption of preventive measures, leaders establish a foundation upon which proactive management practices can flourish. Furthermore, leaders must embrace a culture of learning from failures and near-misses, viewing them not as setbacks but as invaluable opportunities for growth and improvement. This culture of openness and continuous learning fosters an environment where team members feel empowered to voice concerns, propose solutions, and actively contribute to the organization's proactive initiatives.

Training and education serve as essential pillars in supporting the psychological shift toward proactive Kubernetes management. By providing team members with the necessary knowledge, skills, and tools to identify and mitigate potential issues early on, organizations empower their workforce to take proactive action. This empowerment instills a sense of ownership and accountability toward the Kubernetes environment, reinforcing the proactive mindset and fostering a culture of collective responsibility.

Collaboration and effective communication are also integral aspects of the psychological dimension of proactive Kubernetes management. A culture that encourages open sharing of knowledge, experiences, and insights facilitates the dissemination of the proactive mindset throughout the organization. Regular meetings, brainstorming sessions, and knowledge-sharing forums provide opportunities for teams to discuss potential risks, share lessons learned from past incidents, and devise preventive strategies collaboratively. This collaborative approach ensures that everyone is aligned and focused on the importance of proactive management practices.

Recognizing and rewarding proactive behaviors further reinforces their value within the organization. When team members see that their efforts to anticipate and prevent issues are acknowledged and appreciated, it not only boosts morale but also catalyzes further proactive engagement. Recognition can take various forms, including public commendations, performance incentives, or opportunities for professional development, all of which contribute to cultivating a culture that values foresight, vigilance, and a commitment to continuous improvement.

Addressing the psychological aspect of proactive Kubernetes management entails creating an organizational environment that supports and nurtures these proactive behaviors. By recognizing and leveraging the human factors that underpin effective management practices, organizations can enhance their overall resilience, operational efficiency, and adaptability in the face of evolving challenges within cloud-native environments.

# The influence of Kubernetes versioning on strategic planning

The impact of Kubernetes versioning on strategic planning holds immense significance, especially as organizations grapple with the intricacies of maintaining and upgrading their Kubernetes environments. With Kubernetes releasing new versions frequently, each bringing enhancements, bug fixes, and occasionally deprecations of existing features, the landscape of version management presents both challenges and opportunities that demand a proactive approach.

Strategic planning concerning Kubernetes versioning entails several crucial considerations. Foremost among these is the imperative to remain abreast of the release schedule and the contents of upcoming versions. This knowledge empowers organizations to anticipate changes that might impact their deployments and plan upgrades accordingly. It's not merely about embracing new features; it's also about discerning which bug fixes or security patches are indispensable for operational continuity.

Compatibility stands as another pivotal aspect. With each new version, Kubernetes may introduce changes that lack backward compatibility with earlier iterations. Organizations must assess their current deployments, which encompass applications and integrations, to ensure seamless functionality post-upgrade. This evaluation often necessitates rigorous testing in a staging environment before implementing updates in production – a critical step in the strategic planning process.

Employing a phased approach to upgrades can mitigate risks effectively. Rather than overhauling all clusters simultaneously, organizations can commence with non-critical environments to identify potential issues in a controlled setting. This sequential strategy permits adjustments and fine-tuning before advancing to more mission-critical deployments. Planning for these phases, including allocating resources and estimating timeframes, is paramount for minimizing operational disruptions.

Strategic planning also encompasses considering the support life cycle of Kubernetes versions. Each version undergoes a predetermined period during which it receives updates, after which it reaches **end-of-life** (**EOL**) status. Operating on an EOL version can expose organizations to security vulnerabilities and compliance challenges. Hence, planning must encompass timelines for migrating to supported versions to uphold ongoing security and stability.

Furthermore, decisions regarding the adoption of new Kubernetes features or the deprecation of outdated ones demand meticulous deliberation on their potential impact on applications and workflows. While new features offer significant benefits in terms of efficiency, security, and functionality, their integration into existing deployments necessitates careful planning to avoid operational disruptions. Similarly, when features face deprecation, organizations must identify and implement alternatives, which may entail substantial adjustments to their applications or infrastructure.

The influence of Kubernetes versioning on strategic planning is multi-faceted, spanning upgrade paths, compatibility testing, feature adoption, and compliance considerations. By incorporating version management into their strategic planning processes, organizations can uphold the security, stability, and alignment of their Kubernetes environments with operational objectives. This proactive approach empowers them to harness the full potential of Kubernetes while mitigating risks associated with its rapid development pace.

## Emphasizing sustainability in Kubernetes architecture design

Emphasizing sustainability in Kubernetes architecture design has emerged as a paramount concern for organizations striving to develop systems that not only boast efficiency and scalability but also exhibit environmental consciousness and long-term cost-effectiveness. The core objective of sustainable Kubernetes architecture revolves around optimizing resource utilization, curbing wastage, and ensuring infrastructure adaptability to future demands without triggering disproportionate spikes in energy consumption or operational expenditures.

Efficiency stands as the cornerstone of sustainable Kubernetes architecture design. It entails meticulously selecting the appropriate size and type of resources for specific workloads to minimize idle capacities. Over-provisioning resources can result in needless costs and energy consumption, while under-provisioning may compromise performance and reliability. Leveraging practices such as horizontal pod autoscaling, cluster autoscaling, and effective resource management mechanisms facilitate the attainment of a harmonious balance.

Embracing green computing practices within the Kubernetes ecosystem is imperative for sustainability. This involves aligning with cloud providers and data centers that prioritize renewable energy sources and demonstrate a commitment to reducing their carbon footprint. By opting for environmentally conscious hosting solutions, organizations contribute significantly to mitigating their overall environmental impact while fostering sustainability within their technological infrastructure.

Addressing waste reduction is another pivotal facet of sustainable architecture design. Waste within a Kubernetes environment can manifest in various forms, including underutilized resources, redundant application deployments, and excessive logging. By implementing stringent policies for efficient resource allocation, optimized data management strategies, and streamlined application life cycle management practices, organizations can significantly minimize waste and enhance overall sustainability.

Furthermore, designing for adaptability and future growth is paramount to ensuring the long-term sustainability of Kubernetes architecture. This entails crafting systems that possess inherent flexibility to evolve alongside shifting business needs and technological advancements without necessitating complete architectural overhauls. Embracing modular and microservices-based design paradigms facilitates architectural agility, enabling seamless updates, scaling operations, or reconfigurations with minimal disruptions and wastage.

Prioritizing sustainability in Kubernetes architecture design demands a comprehensive approach that transcends immediate organizational objectives to encompass the broader impact on the environment, operational expenditures, and adaptability to future challenges. By emphasizing efficiency optimization, waste reduction, green computing, and architectural flexibility, organizations can construct Kubernetes architectures that not only align with sustainability objectives but also epitomize responsible technology utilization in today's rapidly evolving digital landscape.

## Adaptive leadership for Kubernetes teams

Adaptive leadership stands as a cornerstone in steering Kubernetes teams through the intricate and swift changes inherent in managing cloud-native technologies. This leadership style is particularly suited to environments characterized by frequent technological shifts, where teams must incessantly learn, innovate, and adapt their strategies to uphold effective operations and deliver value.

Leaders who embody an adaptive approach comprehend the significance of flexibility and readiness to adjust strategies based on evolving circumstances and new insights. They acknowledge that in the dynamic realm of Kubernetes, rigid adherence to a singular plan, devoid of consideration for changes in the technology landscape or organizational objectives, can impede progress and innovation.

Central to adaptive leadership is the empowerment of team members. Leaders equip individuals with the tools, resources, and autonomy necessary to make decisions and take initiative. This empowerment fosters a sense of ownership and accountability among team members, compelling them to proactively seek solutions and innovate within their roles. Additionally, it cultivates a culture where learning from mistakes is embraced as a pathway to improvement rather than a cause for reproach.

Effective communication serves as another cornerstone of adaptive leadership. Leaders maintain open channels of communication with their teams, encouraging feedback and sharing insights to anticipate challenges and identify opportunities. This reciprocal communication ensures that team members feel heard and valued, thereby enhancing morale and engagement.

Moreover, adaptive leaders are committed to nurturing a culture of continuous learning. They acknowledge that skills and knowledge that are relevant today may prove inadequate tomorrow. Therefore, they invest in ongoing education and professional development for their team members, enabling them to stay abreast of the latest Kubernetes features, best practices, and industry trends. This dedication to learning ensures that the team remains agile and capable of tackling new challenges as they arise.

Collaboration is also emphasized by adaptive leaders. They recognize that complex issues often necessitate diverse perspectives and expertise to resolve. By fostering collaboration within the team and with external stakeholders, they harness collective knowledge and creativity to devise innovative solutions. This collaborative approach not only enhances problem-solving capabilities but also fortifies team cohesion and resilience.

In managing Kubernetes environments, adaptive leaders prioritize sustainability and adopt a long-term perspective. They strike a balance between immediate operational exigencies and the organization's future vision, ensuring that decisions made today do not compromise the team's adaptability and prosperity in the future. This forward-thinking outlook is indispensable for navigating the uncertainties and opportunities inherent in technological advancement.

Therefore, adaptive leadership transcends mere management of change; it embraces change as an avenue for growth and learning. By leading with flexibility, empowerment, communication, a commitment to learning, collaboration, and a focus on the future, leaders can steer their Kubernetes teams toward excellence in an ever-evolving technological landscape.

# Assessing and anticipating potential pitfalls

This section covers identifying and preparing for Kubernetes challenges, including risk assessments, predictive analytics, capacity planning, and scenario planning. It highlights stress testing, dependency management for stability, and advanced threat modeling for security.

## Conducting thorough Kubernetes risk assessments

Conducting thorough risk assessments within Kubernetes environments is a crucial step in safeguarding the stability, security, and performance of the system. This process entails a methodical examination of the Kubernetes infrastructure, applications, and operational procedures to unearth potential vulnerabilities before they escalate into critical incidents.

To initiate a comprehensive risk assessment for Kubernetes, the first step involves mapping out the entire infrastructure. This includes delineating clusters, nodes, pods, and services to gain a holistic understanding of the system's components and their interdependencies. This comprehensive mapping provides a foundational overview that aids in pinpointing potential vulnerabilities effectively.

After infrastructure mapping, it becomes imperative to evaluate the configuration of the Kubernetes environment. Misconfigurations often serve as a primary source of risk, potentially leading to unauthorized access, data breaches, or service disruptions. Regularly validating configurations against industry best practices and security standards can help you identify and rectify potential issues proactively.

It is essential to assess security vulnerabilities inherent in the Kubernetes ecosystem that encompass both the core platform and any third-party integrations. Staying informed about known vulnerabilities and promptly applying patches and updates is paramount to maintaining the security integrity of the environment. Automated vulnerability scanning tools play a crucial role in continuously monitoring for new risks and ensuring timely remediation.

The risk assessment process involves scrutinizing deployment processes and practices. This includes evaluating CI/CD pipelines, container image management practices, and deployment strategies for potential risks. Ensuring the use of trusted images, implementing robust gating and approval processes in the CI/CD pipeline, and adopting strategies such as blue-green or canary deployments mitigate the impact of issues during updates.

Another critical aspect of risk assessment is analyzing the resilience and recovery capabilities of the Kubernetes environment. This entails evaluating disaster recovery plans, backup strategies, and the system's ability to quickly recover from incidents. Understanding how the system behaves under failure conditions and implementing robust mechanisms for restoration significantly reduces the risk of prolonged outages or data loss.

By diligently assessing risks within Kubernetes environments, organizations can proactively address vulnerabilities, fortify their security posture, and enhance the reliability and performance of their deployments. This proactive approach to risk management cultivates a more resilient and secure

Kubernetes ecosystem, empowering organizations to leverage the benefits of cloud-native technologies with confidence and assurance.

## Utilizing predictive analytics in Kubernetes environments

Employing predictive analytics within Kubernetes environments offers a potent method for anticipating potential issues and enhancing system performance before they escalate. This strategy involves scrutinizing data from diverse sources within the Kubernetes ecosystem to uncover patterns, trends, and anomalies that may signal future risks or areas for improvement.

At the core of predictive analytics in Kubernetes lies collecting comprehensive data. This encompasses metrics related to resource usage, such as CPU, memory, and storage, alongside operational data such as container start times, failure rates, and network traffic patterns. By amassing a wealth of data, teams can construct a detailed overview of the system's behavior and performance.

Once data collection is underway, the subsequent step involves leveraging analytics tools and machine learning models to scrutinize the data. These tools can unearth correlations and patterns that might elude immediate notice. For instance, a model could forecast an impending surge in traffic for a specific service based on historical trends, empowering teams to preemptively adjust resources to meet the surge in demand.

Predictive analytics can also aid in detecting potential security threats. By scrutinizing patterns of network traffic and access logs, the system can flag unusual activity that may signify a security breach or exploitation of a vulnerability. Early detection equips teams to swiftly respond and mitigate the threat.

Another valuable application of predictive analytics lies in capacity planning. By deciphering trends in resource usage and application performance, organizations can make informed decisions about when to scale resources or invest in infrastructure upgrades. This proactive approach ensures that the system remains responsive and efficient without unnecessarily provisioning resources.

Furthermore, predictive analytics contributes to effective cost management. By forecasting future resource requirements, organizations can optimize spending on cloud resources, circumventing unnecessary costs while ensuring that performance and availability objectives are met.

Implementing predictive analytics in Kubernetes demands a blend of technical tools and expertise. Teams require proficiency in data science and machine learning, coupled with a profound comprehension of the Kubernetes platform and its hosted applications. Moreover, integrating predictive analytics into operational processes necessitates a cultural shift toward data-driven decision-making.

Leveraging predictive analytics in Kubernetes environments empowers teams to transition from reactive to proactive management. By analyzing data to anticipate future scenarios, organizations can enhance performance, bolster security, plan capacity more efficiently, and manage costs effectively. This forward-thinking approach fosters more resilient, efficient, and cost-effective Kubernetes operations.

## The art of Kubernetes capacity planning

The art of capacity planning in Kubernetes environments involves carefully balancing resource allocation to meet current and future demands without over or underutilizing infrastructure. This process is crucial for maintaining optimal performance and efficiency, ensuring that applications run smoothly while keeping costs in check.

At the heart of effective capacity planning is a deep understanding of the workloads running in the Kubernetes cluster. This includes knowing the resource requirements of each application, such as CPU, memory, and storage, and how these requirements change under different loads. Monitoring these metrics over time provides valuable insights into usage patterns and growth trends, which are essential for forecasting future needs.

An important aspect of capacity planning is the implementation of resource requests and limits for pods in Kubernetes. These settings allow administrators to specify the minimum and maximum amount of resources each pod can use, helping to prevent any single application from consuming more than its fair share of resources. By carefully configuring these parameters, teams can ensure that all applications have the resources they need to perform optimally while maximizing the overall utilization of the cluster.

Scalability is another key consideration in capacity planning. Kubernetes offers several mechanisms to automatically scale resources, including horizontal pod autoscaling, which adjusts the number of pod replicas based on demand, and cluster autoscaling, which adds or removes nodes from the cluster. Leveraging these features allows organizations to respond dynamically to changes in workload, ensuring that the environment can handle peak loads without maintaining excess capacity.

Effective capacity planning also requires collaboration between development and operations teams. Developers need to provide accurate estimates of their applications' resource needs and understand how changes in their code might impact resource consumption. Operations teams, on the other hand, should share insights into the overall capacity and performance of the Kubernetes environment, helping to guide development practices.

Cost management is an integral part of capacity planning, especially for organizations using cloud-based Kubernetes services. By aligning resource allocation with actual usage, teams can avoid unnecessary expenses associated with overprovisioning. Tools and platforms that offer cost analysis and forecasting can help organizations make more informed decisions about their Kubernetes infrastructure investments.

Regularly reviewing and adjusting capacity plans is essential because the needs of applications and the availability of resources change over time. This ongoing process involves revisiting resource allocations, scaling settings, and cost projections to ensure that the Kubernetes environment remains aligned with organizational goals and requirements.

In practice, mastering the art of capacity planning in Kubernetes is about combining detailed knowledge of application needs with the flexibility offered by Kubernetes' scaling features. It requires a proactive approach to monitoring, a commitment to collaboration across teams, and a continuous effort to optimize resource use and costs. By achieving this balance, organizations can ensure their Kubernetes environments are both powerful and cost-effective, ready to support their applications now and in the future.

## Scenario planning – preparing for the unexpected

Planning for potential scenarios within Kubernetes environments is a strategic method that helps organizations deal with unexpected situations effectively. This approach involves imagining different future situations, both likely and unlikely, to create plans that ensure the resilience and continuity of Kubernetes operations under various conditions.

The process starts with identifying a range of potential scenarios that could impact the Kubernetes environment. These scenarios might include sudden spikes in demand, infrastructure failures, security breaches, or significant changes in operational costs. By considering a broad spectrum of possibilities, from the mundane to the catastrophic, teams can better prepare for the unexpected.

For each scenario, the next step is to assess the potential impact on the Kubernetes infrastructure and the applications it supports. This involves analyzing how different events might affect resource utilization, application performance, and overall system stability. Understanding these impacts helps in prioritizing which scenarios require more immediate attention and resources to mitigate.

Developing response strategies for each identified scenario is crucial. These strategies should outline specific actions to be taken in response to events, including scaling resources, rerouting traffic, applying security patches, or executing disaster recovery plans. By having these plans in place, teams can respond swiftly and effectively, minimizing downtime and reducing the negative impact on operations.

Testing and simulation play a key role in scenario planning. Conducting drills or using simulation tools to mimic the scenarios can reveal weaknesses in the response plans and provide valuable insights into how the Kubernetes environment behaves under stress. This hands-on practice helps teams refine their strategies and ensures they are prepared to act in real-world situations.

Incorporating flexibility into the Kubernetes architecture is another important aspect of scenario planning. Designing systems that can adapt to changing conditions, such as using microservices architectures or implementing automated scaling, enhances the ability to respond to various scenarios without significant manual intervention.

Effective communication and documentation are essential components of scenario planning. Ensuring that all team members understand the plans and know their roles in responding to different scenarios fosters a cohesive and coordinated effort when unexpected events occur. Detailed documentation of the scenarios, impact assessments, and response strategies also facilitates ongoing review and updates to the plans as conditions change.

By engaging in scenario planning, organizations can establish a proactive and resilient approach to managing their Kubernetes environments. This preparation helps mitigate risks associated with unexpected events and positions teams to navigate challenges confidently, ensuring the continuity and stability of their operations.

## Stress testing your Kubernetes infrastructure

Testing the resilience of your Kubernetes infrastructure through stress testing is crucial. It involves creating scenarios of high loads to see how the system handles pressure. This helps reveal the infrastructure's capacity limits, find any bottlenecks, and ensure that the system remains stable and performs well during peak usage times.

The first step in stress testing is to define the goals and parameters of the test. This includes determining the metrics to measure, such as response times, throughput, and resource utilization, and setting the thresholds for acceptable performance under stress. Clear objectives help focus the testing efforts and provide benchmarks against which to evaluate the results.

Selecting the right tools and frameworks for stress testing is essential. There are a variety of open source and commercial tools available that can generate high volumes of traffic and simulate various types of workloads on the Kubernetes cluster. Choosing tools that align with the specific testing goals and the nature of the applications running in the cluster is crucial for obtaining meaningful results.

Designing the test scenarios is a critical phase. Scenarios should mimic realistic usage patterns as closely as possible, including the mix of read and write operations, user interactions, and API calls that the applications typically handle. Incorporating a range of scenarios, from expected peak loads to extreme conditions, ensures a comprehensive assessment of the system's resilience.

Executing stress tests requires careful planning and monitoring. It's important to gradually increase the load on the system, monitoring the performance and behavior of the Kubernetes infrastructure and the applications it hosts. This approach helps identify at what point the system begins to degrade and which components are the first to fail under stress.

Analyzing the results of the stress tests involves examining the collected metrics to identify any performance issues, resource constraints, or failures. This analysis should provide insights into how the system behaves under different load levels, where bottlenecks occur, and what the maximum capacity of the infrastructure is before it becomes unstable.

Based on the findings, the next step may involve making adjustments to the Kubernetes configuration, such as tuning resource allocations, scaling out components, or optimizing application code. The goal is to address the issues that were identified and improve the system's ability to handle high load conditions.

It's also important to document the stress testing process, including the test scenarios, execution details, results, and any actions taken in response to the findings. This documentation serves as a valuable reference for future testing and for understanding the system's performance characteristics.

By stress testing your Kubernetes infrastructure, you gain valuable insights into the system's performance limits and resilience. This knowledge enables you to make informed decisions about scaling, resource allocation, and architectural improvements, ensuring that the infrastructure can support the demands of your applications even under the most challenging conditions.

# Dependency management and its impact on stability

Managing dependencies in Kubernetes environments is crucial for ensuring the stability and reliability of applications. Dependencies in this context refer to the various software libraries, packages, and external services that applications need to function. If they're not managed properly, these dependencies can introduce vulnerabilities, compatibility issues, or unexpected behavior, impacting the overall stability of the Kubernetes infrastructure.

The first step in effective dependency management is to create a comprehensive inventory of all dependencies for each application running in Kubernetes. This inventory should include details about the version, source, and purpose of each dependency. Having a clear understanding of what dependencies exist enables teams to monitor them for updates, vulnerabilities, and deprecations more effectively.

Regularly updating dependencies is essential for maintaining security and functionality. However, updates must be approached with caution to avoid introducing breaking changes. Automated tools can help identify new versions of dependencies, but thorough testing is necessary to ensure that updates do not adversely affect the application. Implementing a robust CI/CD pipeline that includes automated testing can streamline this process, allowing teams to update dependencies with confidence.

Dependency isolation is another important strategy. By isolating dependencies as much as possible, teams can prevent conflicts between different versions of the same library that are used by multiple applications. Containerization inherently supports dependency isolation by allowing each application to include its specific dependencies in its container image. Further isolation can be achieved through the use of Kubernetes namespaces, which separate resources within the same cluster.

Managing transitive dependencies – those not directly included by the application but required by its dependencies – also requires attention. These can be particularly challenging to track and update. Tools that provide dependency tree analysis can help teams understand and manage these indirect dependencies, ensuring that they do not introduce security vulnerabilities or stability issues.

Monitoring for vulnerabilities in dependencies is critical. Leveraging security scanning tools that can identify known vulnerabilities in dependencies helps teams take proactive measures to mitigate risks. These tools can be integrated into the CI/CD pipeline to automatically scan for vulnerabilities whenever changes are made, ensuring ongoing vigilance.

Version pinning is a practice that can enhance stability by specifying exact versions of dependencies to use, rather than relying on the latest versions. While this approach can prevent unexpected changes, it also requires teams to manually update these versions to benefit from bug fixes and security patches. Balancing the use of version pinning with regular updates is key to maintaining both stability and security.

Documenting dependency management policies and practices ensures that all team members understand how to handle dependencies consistently. This documentation should include guidelines for adding new dependencies, updating existing ones, and responding to security vulnerabilities.

Effective dependency management in Kubernetes environments involves carefully tracking, updating, and isolating dependencies to prevent issues that could destabilize the infrastructure. By implementing best practices for dependency management, teams can ensure that their applications remain secure, functional, and reliable, supporting the overall stability of the Kubernetes ecosystem.

## Advanced Kubernetes threat modeling for security

Advanced threat modeling for security in Kubernetes environments is a systematic approach to identifying and addressing potential security threats. This process involves understanding the specific components of the Kubernetes architecture, how they interact, and where vulnerabilities might exist. By anticipating potential attack vectors, organizations can implement more robust security measures to protect their Kubernetes deployments.

The first step in threat modeling is to create a detailed map of the Kubernetes environment. This includes not just the Kubernetes clusters, but also the underlying infrastructure, associated services, and connections to external systems. Mapping out these components helps in identifying where sensitive data is stored, how it moves within the system, and which parts of the architecture are exposed to potential attackers.

Once the environment has been mapped, the next step is to identify potential threats. This involves considering a wide range of possible attack scenarios, including but not limited to unauthorized access, data breaches, denial-of-service attacks, and insider threats. Each identified threat is analyzed in terms of its likelihood and potential impact, allowing organizations to prioritize their security efforts.

For each potential threat, teams must then identify existing security controls and assess their effectiveness. This includes not only Kubernetes-native security features, such as RBAC and network policies, but also additional security measures that are implemented by the organization, such as firewalls, intrusion detection systems, and security monitoring tools.

Identifying gaps in the existing security posture is a critical outcome of the threat modeling process. This may involve recognizing areas where security controls are lacking, configurations are suboptimal, or additional monitoring is needed. Addressing these gaps requires a combination of updating configurations, implementing additional security measures, and enhancing monitoring and alerting capabilities.

An important aspect of advanced threat modeling is considering the evolving threat landscape. As new vulnerabilities are discovered and attack techniques evolve, the threat model must be updated to reflect these changes. Regularly reviewing and updating the threat model ensures that the Kubernetes environment remains protected against emerging threats.

Engaging in simulations and red team exercises can also enhance the effectiveness of threat modeling. By simulating attacks or conducting penetration testing, organizations can validate their threat models and the effectiveness of their security controls in a controlled environment. These exercises can reveal previously unidentified vulnerabilities and provide valuable insights into how to strengthen security.

As the environment changes, with new deployments, updated configurations, or evolving business requirements, the threat model must be revisited and revised. Continuous improvement of the security posture, guided by the threat modeling process, helps organizations stay ahead of potential security threats.

Employing advanced threat modeling for security in Kubernetes environments is a proactive and organized method for pinpointing and addressing potential security risks. By comprehensively understanding the environment, recognizing potential threats, evaluating existing controls, and making ongoing updates to the threat model, organizations can substantially improve the security and robustness of their Kubernetes deployments.

# Implementing preventative measures

This section outlines implementing preventive measures in Kubernetes and emphasizes documentation, **standard operating procedures** (**SOPs**), disaster simulation frameworks, financial governance, policy enforcement, audit and compliance programs, and strategic planning for capacity and growth.

## The importance of documentation and SOPs

Thorough documentation and established SOPs are indispensable in Kubernetes environments. These fundamental components are essential for preserving system integrity, promoting operational uniformity, and facilitating prompt responses to incidents. Documentation and SOPs form the backbone for teams navigating the intricate terrain of Kubernetes management, offering clear direction and a dependable reference for both routine tasks and urgent scenarios.

Documentation in Kubernetes contexts covers a wide array of information, including architectural overviews, configuration details, deployment processes, and troubleshooting guides. Having comprehensive documentation ensures that all team members, from developers to operations personnel, have access to the information they need to understand the environment fully. This is especially valuable in onboarding new team members, reducing the learning curve, and helping them become productive more quickly.

SOPs complement documentation by providing step-by-step instructions for routine tasks and response strategies for potential issues. SOPs ensure that operations are performed consistently, reducing the likelihood of human error and variability in system management. Whether it's deploying a new service, scaling resources, or responding to a security alert, SOPs offer a structured approach that guides team members through the process.

One of the key benefits of well-crafted documentation and SOPs is the facilitation of knowledge sharing within the organization. Instead of relying on tacit knowledge held by a few individuals, documented procedures ensure that critical operational knowledge is accessible to everyone. This democratization of knowledge not only enhances team efficiency but also mitigates the risk associated with personnel changes.

In the context of incident response, documentation and SOPs become even more critical. When faced with system outages or security breaches, having predefined procedures allows teams to act swiftly and effectively. SOPs for incident response should detail communication protocols, escalation paths, and remediation steps, ensuring a coordinated and comprehensive response to minimize downtime and mitigate impacts.

To maintain their relevance and effectiveness, documentation and SOPs require regular review and updates. As the Kubernetes environment evolves, with changes in infrastructure, applications, and operational practices, the associated documentation and procedures must be revised to reflect the current state. This ongoing maintenance ensures that the information remains accurate and useful, supporting the organization's operational needs.

Implementing preventive measures in Kubernetes environments is significantly enhanced by the presence of robust documentation and well-defined SOPs. These tools not only improve operational efficiency and consistency but also strengthen the organization's ability to respond to challenges proactively. By investing in the development and maintenance of documentation and SOPs, organizations can ensure a solid foundation for secure, reliable, and efficient Kubernetes operations.

## Building a Kubernetes disaster simulation framework

Developing a disaster simulation framework within Kubernetes environments is crucial for preparing and assessing the resilience of systems against possible failures or catastrophic events. This proactive strategy entails crafting scenarios that replicate real-life disasters, spanning from minor interruptions to significant outages. These simulations enable teams to gauge the efficacy of their response plans and the resilience of their infrastructure under challenging conditions.

The first step in building such a framework is to identify the range of disasters that could potentially impact the Kubernetes environment. These scenarios might include hardware failures, network disruptions, security breaches, data corruption, or complete data center outages. By considering a wide spectrum of possibilities, organizations can ensure they are prepared for various incidents.

Once potential disaster scenarios have been identified, the next phase involves designing simulation exercises that accurately replicate these conditions within the Kubernetes environment. This requires careful planning to ensure that simulations are realistic yet do not disrupt actual production workloads. Techniques such as chaos engineering, where components are intentionally disrupted to test system resilience, can be particularly effective in this context.

Developing clear objectives for each simulation exercise is crucial. These objectives might include assessing the effectiveness of failover mechanisms, the efficiency of backup and restore procedures, or the response times of operational teams. By setting specific goals, teams can focus their efforts and measure the outcomes of each simulation more accurately.

Implementing the disaster simulation framework requires robust tooling and automation. Tools that can programmatically introduce failures or degrade performance are essential for conducting simulations consistently and repeatably. Automation ensures that simulations can be run regularly without requiring excessive manual intervention, making it easier to integrate disaster preparedness into the regular operational routine.

Training and involving the entire team in disaster simulation exercises is vital. These simulations provide invaluable learning opportunities, allowing team members to practice their responses to various scenarios in a controlled environment. This hands-on experience is crucial for building confidence and ensuring that everyone knows their role during an actual disaster.

Documenting the outcomes of each simulation exercise is another critical component. This documentation should include details of the scenario, the actions taken in response, the outcomes, and any lessons learned. Reviewing this documentation allows teams to refine their disaster response strategies, update SOPs, and make necessary adjustments to the Kubernetes configuration or architecture to enhance resilience.

Regular review and iteration of the disaster simulation framework are essential. As the Kubernetes environment evolves, so too should the disaster preparedness strategies. Regularly updating and expanding the simulation scenarios to reflect new risks and changes in the infrastructure ensures that the organization remains well-prepared for any eventuality.

Building a Kubernetes disaster simulation framework is a proactive measure that significantly enhances an organization's preparedness for unexpected events. By systematically identifying potential disasters, designing realistic simulations, and engaging the entire team in these exercises, organizations can ensure that their Kubernetes environments are resilient, responsive, and capable of withstanding the challenges of the real world.

## Implementing Kubernetes financial governance to avoid cost surprises

Implementing financial governance within Kubernetes environments is pivotal to managing costs effectively and preventing unexpected expenses. Kubernetes, with its dynamic scaling capabilities and complex resource allocation models, can lead to significant cost variations if not monitored and managed carefully. Financial governance involves setting budgets, monitoring resource utilization, and optimizing deployments to align with financial objectives without compromising performance or reliability.

The foundation of effective Kubernetes financial governance is establishing clear budgetary constraints for different teams or projects. By allocating specific budgets, organizations can ensure that resource usage stays within predefined limits, preventing cost overruns. These budgets should be based on detailed analyses of past usage patterns, projected needs, and organizational financial goals, allowing for a balance between operational flexibility and cost control.

Monitoring and tracking resource utilization in real time is another critical aspect of financial governance. This requires implementing tooling that provides visibility into how resources are being consumed across different Kubernetes clusters and workloads. With detailed insights into resource usage, teams can identify areas where efficiencies can be gained, such as underutilized resources that can be scaled down or more cost-effective resource types that can be utilized.

Cost optimization strategies must also be incorporated into the Kubernetes operational model. This includes selecting the right mix of resource types and sizes, leveraging reserved instances or spot instances where appropriate, and implementing auto-scaling to adjust resource allocation dynamically based on demand. Additionally, adopting practices such as container right-sizing and efficient container image management can further reduce unnecessary resource consumption and associated costs.

Establishing policies and procedures for financial governance is essential to ensure that cost management practices are applied consistently across the organization. These policies might cover aspects such as approval processes for resource allocation increases, guidelines for using cloud provider services, and procedures for addressing budget overruns. Regular training and communication efforts can help reinforce the importance of financial governance and ensure that all team members understand their roles in managing costs.

Regular reviews and audits of Kubernetes spending are crucial for maintaining effective financial governance. These reviews can uncover unexpected cost drivers, assess the effectiveness of cost optimization measures, and identify opportunities for further savings. Feedback from these reviews can then be used to refine budgetary allocations, adjust monitoring and optimization strategies, and update governance policies as needed.

Involving stakeholders from across the organization, including finance, operations, and development teams, in the financial governance process ensures a holistic approach to managing Kubernetes costs. Collaboration between these groups can lead to more informed decision-making, with financial considerations balanced against operational requirements and development goals.

Kubernetes financial governance is a comprehensive approach to managing costs and preventing budgetary surprises. By establishing clear budgets, monitoring resource utilization, optimizing deployments, implementing governance policies, conducting regular reviews, and fostering collaboration across the organization, teams can ensure that their Kubernetes environments are both cost-effective and aligned with broader financial objectives. This proactive approach to financial governance supports sustainable growth and operational efficiency in Kubernetes deployments.

## Kubernetes policy enforcement mechanisms

Deploying policy enforcement mechanisms in Kubernetes environments is crucial to upholding operational integrity, security, and compliance. These mechanisms furnish the requisite controls to guarantee that deployments and configurations align with organizational norms and optimal practices. By automating policy enforcement, organizations can avert the deployment of non-compliant resources, thereby mitigating the risk of security breaches, operational interruptions, and compliance breaches.

Kubernetes offers several native features and third-party tools that are designed for policy enforcement. One of the core Kubernetes features for policy enforcement is RBAC. This allows administrators to define roles with specific permissions and assign these roles to users, groups, or service accounts. This ensures that only authorized personnel have access to perform certain actions within the Kubernetes environment, such as creating or modifying resources.

Another powerful policy enforcement tool is **PodSecurityPolicy** (**PSP**), which controls the security specifications a pod must adhere to for it to be accepted into the system. PSP can enforce various security-related policies, including running containers as non-root users, preventing privilege escalation, and controlling access to host filesystems and networks. Although PSP is being deprecated in favor of newer solutions, its concept remains a critical aspect of Kubernetes security.

Network policies in Kubernetes allow administrators to control the flow of traffic between pod groups. By defining network policies, teams can enforce rules about which pods can communicate with each other, thereby limiting the potential attack surface for network-based threats.

For more advanced and customizable policy enforcement, organizations often turn to third-party tools such as **Open Policy Agent** (**OPA**) and Kyverno. These tools integrate with Kubernetes to provide a policy-as-code approach, allowing administrators to define policies in a declarative manner. Policies can cover a wide range of requirements, from resource naming conventions and image registry restrictions to minimum resource allocations and maximum limits.

Implementing policy enforcement mechanisms also involves setting up validation and auditing processes. Admission controllers in Kubernetes, for example, can be used to review and approve or reject requests to create or update resources based on defined policies. Auditing mechanisms, meanwhile, can track and log all actions taken in the environment, providing an audit trail that can be reviewed for compliance and operational analysis.

To ensure the effectiveness of policy enforcement mechanisms, organizations must also invest in training and awareness for their development and operations teams. Educating team members on the importance of compliance and how to work within the boundaries of enforced policies fosters a culture of security and operational excellence.

Regularly reviewing and updating policies is crucial as organizational needs and external requirements evolve. This continuous improvement process ensures that policy enforcement mechanisms remain relevant and effective in addressing new challenges and regulatory requirements.

Policy enforcement mechanisms in Kubernetes are vital for securing the environment, ensuring operational consistency, and maintaining compliance with internal and external standards. By leveraging Kubernetes features and third-party tools for policy enforcement, along with implementing robust validation, auditing, and education practices, organizations can create a secure and compliant Kubernetes ecosystem that supports their operational objectives.

## Kubernetes audit and compliance program

Establishing a Kubernetes audit and compliance program is essential for organizations to ensure that their Kubernetes environments adhere to internal policies, industry standards, and regulatory requirements. This program involves systematically reviewing and verifying all aspects of the Kubernetes infrastructure, including configurations, access controls, network policies, and resource usage. By identifying non-compliance issues and security gaps, organizations can take corrective actions to mitigate risks and maintain the integrity of their Kubernetes deployments.

The first step in creating an audit and compliance program is to define the compliance requirements that are specific to the organization's industry and operational context. This could include regulations such as GDPR for data protection, HIPAA for healthcare information, or PCI DSS for payment card data. Understanding these requirements is crucial for developing a comprehensive audit framework that addresses all relevant compliance aspects.

Once the compliance requirements have been established, the next phase involves developing an audit checklist or framework that outlines the specific items to be reviewed. This checklist should cover various areas, including but not limited to Kubernetes cluster configurations, network policies, RBAC, logging and monitoring practices, and data storage and protection mechanisms. The checklist serves as a guide for the audit process, ensuring a thorough and systematic review.

Implementing tooling for continuous monitoring and auditing is another critical component of the program. Tools that can automatically scan Kubernetes configurations, detect deviations from best practices, and identify potential security vulnerabilities are invaluable for maintaining ongoing compliance. These tools can provide real-time alerts about non-compliance issues or security threats, allowing for prompt remediation.

Conducting regular audits according to the established framework is essential for the effectiveness of the compliance program. These audits can be performed internally by the organization's security or compliance teams, or by external auditors for an independent review. Regular audits help to identify compliance gaps, assess the effectiveness of current controls, and highlight areas for improvement.

Documenting the findings and actions taken during the audits is crucial for accountability and continuous improvement. Audit reports should detail any non-compliance issues identified, the risks associated with these issues, and the corrective actions taken or recommended. These reports not only serve as a record of compliance efforts but also as a basis for refining the audit and compliance program over time.

Training and awareness are also vital components of a successful audit and compliance program. Ensuring that all team members understand the compliance requirements, the importance of adhering to best practices, and their roles in maintaining compliance helps foster a culture of security and compliance within the organization.

The audit and compliance program should be reviewed and updated regularly to reflect changes in regulatory requirements, organizational policies, and the Kubernetes environment itself. This iterative process ensures that the program remains relevant and effective in addressing new challenges and compliance obligations.

A Kubernetes audit and compliance program is a comprehensive approach to ensuring that Kubernetes environments meet the required security standards and regulatory requirements. Through systematic audits, continuous monitoring, documentation, training, and regular updates, organizations can mitigate risks, enhance security, and ensure compliance in their Kubernetes deployments.

## Kubernetes capacity and growth strategic planning

Strategic planning for Kubernetes' capacity and growth involves anticipating the future needs of your applications and infrastructure to ensure they can scale effectively without incurring unnecessary costs or performance bottlenecks. This planning is crucial for organizations that rely on Kubernetes for deploying and managing their applications as it impacts both operational efficiency and the ability to meet business objectives.

The process begins with a thorough analysis of current resource usage patterns and trends within the Kubernetes environment. By examining metrics such as CPU, memory, storage utilization, and network traffic over time, organizations can identify usage patterns and predict future demands. This analysis should take into account not only the current workloads but also planned projects and potential business growth that could increase demand for resources.

Effective capacity and growth planning also requires understanding the scalability limits of your Kubernetes clusters and the underlying infrastructure. This includes assessing the maximum capacity of nodes, pods, and services that can be supported within a single cluster and identifying when it might be necessary to add additional clusters or redesign the architecture for better scalability.

Incorporating automation and scalability features of Kubernetes, such as horizontal pod auto scalers, cluster auto scalers, and custom resource definitions, into your strategic planning can enhance the system's ability to adapt to changing demands dynamically. These tools allow Kubernetes to automatically adjust resources based on workload requirements, ensuring that applications have the resources they need when they need them, without overprovisioning.

Cost management is another critical aspect of capacity and growth planning. As Kubernetes environments scale, costs can escalate quickly, especially in cloud-based deployments. Therefore, strategic planning should include budget forecasting and cost optimization strategies, such as selecting the right mix of on-demand and reserved instances. It should also involve implementing cost monitoring and alerting tools to keep expenses under control.

Collaboration across teams is essential for effective capacity and growth planning. Development, operations, finance, and business units should all contribute to the planning process, ensuring that technical capacity aligns with business goals and financial constraints. This collaborative approach helps to balance operational requirements with business objectives, leading to a more efficient and effective deployment strategy.

Regularly revisiting and updating the capacity and growth plan is necessary to adapt to changes in the business environment, technology advancements, or shifts in user demand. This iterative process ensures that the Kubernetes infrastructure remains aligned with the organization's needs, supporting growth and innovation while managing costs and maintaining performance.

Strategic planning for Kubernetes capacity and growth is a multifaceted process that requires a deep understanding of current and future demands, the scalability of the infrastructure, cost management, and cross-functional collaboration. By adopting a proactive and comprehensive approach to planning, organizations can ensure their Kubernetes environments are prepared to support their operational and business objectives efficiently and effectively.

## Summary

This chapter explored proactive assessment and prevention strategies in Kubernetes environments, stressing the significance of a proactive mindset in management. It highlighted prevention over correction and early detection as fundamental principles. Understanding trends within the Kubernetes ecosystem and the psychological aspects of proactive management were also addressed. Strategic planning considerations, including Kubernetes versioning and sustainability in architecture design, were examined. Additionally, insights into assessing and anticipating potential pitfalls, conducting risk assessments, and using predictive analytics were provided. Practical measures for implementing preventative strategies, such as documentation, disaster simulation frameworks, financial governance, policy enforcement mechanisms, and audit and compliance programs, were outlined. Overall, proactive approaches were emphasized to ensure the health, stability, and security of Kubernetes environments.

In the next chapter, we will consolidate all the elements we have explored throughout this book and draw conclusions from our discussions.

# Bringing It All Together

This chapter synthesizes essential lessons from the entire book, focusing on recognizing Kubernetes anti-patterns, addressing challenges with informed solutions, and embracing best practices for operational excellence. It discusses strategic planning for future-proofing deployments, the impact of architectural choices, and the importance of creating stable environments through resilience, security, simplification, and tool optimization. Additionally, it highlights fostering a culture of continuous improvement, innovation, and the critical role of leadership in Kubernetes strategy, equipping readers for effective Kubernetes management and growth.

We'll cover the following topics in this chapter:

- Summarizing key takeaways from the book

- Applying knowledge to create stable Kubernetes environments

- Encouraging a culture of continuous improvement

## Summarizing key takeaways from the book

This section distills key insights from the book, covering Kubernetes anti-patterns, pivotal challenges and solutions, operational best practices, future-proofing strategies, and the influence of architectural choices on deployments.

### Core concepts of Kubernetes anti-patterns

To master the environment of Kubernetes, it's critical to start at the ground level with a clear understanding of what Kubernetes anti-patterns are. These are essentially misapplied practices or configurations that, despite possibly offering immediate relief or seeming to be the easiest route at first, can lead to larger, more complex problems in your Kubernetes deployments. They emerge from a variety of sources: misconceptions about how Kubernetes operates, misconfigurations due to a lack of detailed knowledge, or even well-intentioned practices that don't scale well or align poorly with Kubernetes' architecture.

The concept of anti-patterns in Kubernetes is not just about identifying what not to do. It's about understanding the why—why certain practices lead to negative outcomes, and why alternatives, though they might require more effort upfront, lead to healthier, more sustainable systems. For instance, one might encounter the anti-pattern of overprovisioning resources to avoid running out, which seems prudent. However, this practice ignores Kubernetes' ability to dynamically manage workloads and resources, leading to inefficiencies and unnecessary costs.

In the same vein, another common anti-pattern is the underutilization of Kubernetes' native monitoring and logging tools. Teams might rely on external tools they're already familiar with or skip detailed monitoring altogether, missing out on critical insights into their application's performance and health that could preempt failures or performance bottlenecks.

Kubernetes environments thrive on best practices—practices that have been honed through countless deployments, failures, and successes across the global Kubernetes community. These include embracing declarative configurations, which ensure systems are reproducible and traceable, or adhering to the **principle of least privilege** (**PoLP**) when configuring access controls, enhancing the security posture of the deployment.

The journey through recognizing and correcting Kubernetes anti-patterns is ongoing. As Kubernetes evolves, so too do the anti-patterns and the best practices for avoiding them. The landscape is dynamic, with new features and capabilities added regularly, each bringing new potential for missteps but also opportunities for enhancement.

For anyone looking to build and maintain Kubernetes deployments, the key takeaways are clear. First, invest time in understanding the foundational concepts and capabilities of Kubernetes to make informed decisions rather than defaulting to familiar or simplistic solutions that may lead to anti-patterns. Second, engage with the broader Kubernetes community—there's a wealth of knowledge and experience to draw from, offering insights into common pitfalls and proven strategies for success. Lastly, adopt a mindset of continuous improvement and always be willing to re-evaluate and adjust practices in response to new information, experiences, and the evolving landscape of Kubernetes itself.

By focusing on these core concepts, practitioners can navigate the complexities of Kubernetes with greater confidence, avoiding common pitfalls that lead to suboptimal deployments and, instead, leveraging the full potential of this powerful tool for container orchestration.

## Key challenges and solutions overview

Navigating Kubernetes presents users with diverse challenges, often appearing formidable. These obstacles span from optimizing resource management to guaranteeing application security and scalability. Over time, the Kubernetes community has crafted and honed myriad solutions for these challenges. A thorough exploration of these hurdles and their corresponding remedies provides indispensable guidance for those immersed in Kubernetes landscapes.

One common challenge is the efficient allocation of resources. Without careful planning, teams can either allocate too many resources, leading to wastage, or too few, leading to performance issues. The solution lies in understanding Kubernetes' resource management features, such as requests and limits, and autoscaling capabilities. These features allow for dynamic adjustment of resources based on actual usage, ensuring applications have what they need without unnecessary expenditure.

Security poses another significant challenge. Protecting applications and data in Kubernetes requires a multifaceted approach. Solutions include implementing RBAC to limit access based on PoLP, using network policies to control traffic flow between pods, and ensuring images are scanned for vulnerabilities before deployment. These practices help create a robust security posture, mitigating risks and protecting against potential breaches.

Scalability is also a crucial aspect of Kubernetes deployments. As applications grow, they must scale to meet increasing demand. Kubernetes offers horizontal pod autoscaling, which adjusts the number of pod replicas based on defined metrics such as CPU usage. However, effectively using these features requires a solid understanding of the underlying application behavior and traffic patterns to configure scaling policies that respond appropriately to demand.

Another challenge lies in monitoring and logging. With the complexity of Kubernetes environments, gaining visibility into application performance and system health is essential. The solution involves leveraging Kubernetes' built-in tools along with third-party solutions to create a comprehensive monitoring and logging strategy. This enables teams to detect and respond to issues promptly, often before they impact users.

The process of addressing these challenges is not static. As Kubernetes continues to evolve, new challenges arise, and the community develops new solutions. Engaging with the community through forums, conferences, and collaborative projects is crucial for staying informed about best practices and emerging trends. This engagement also provides opportunities to share experiences and learn from others, fostering a culture of continuous improvement.

For those navigating the Kubernetes landscape, understanding these key challenges and their solutions is crucial. It provides a foundation for building and maintaining resilient, efficient, and secure deployments. Moreover, it underscores the importance of continuous learning and adaptation, ensuring that Kubernetes environments can meet the demands of today and tomorrow.

## Best practices for operational excellence

Achieving operational excellence in Kubernetes is a goal that demands adherence to a set of best practices. These practices are distilled from the experiences of countless professionals who have navigated the complexities of Kubernetes to find the most efficient, secure, and scalable ways to manage their deployments. Understanding and implementing these best practices can significantly improve the reliability and performance of Kubernetes environments.

One crucial practice is the implementation of CI/CD pipelines. These pipelines automate the process of testing and deploying applications, ensuring that changes are systematically validated before being introduced to production. This reduces the risk of errors and downtime, promoting a more stable operational environment.

Another key practice is the embracement of declarative configurations. By defining the desired state of applications and infrastructure in configuration files, teams can ensure consistency, reproducibility, and automation in deployments. This approach minimizes manual interventions, reducing the potential for human error and making it easier to recover from failures.

Effective resource management is also central to operational excellence in Kubernetes. This involves setting appropriate requests and limits for resources such as CPU and memory, preventing any single application from consuming disproportionate resources that could impact overall system stability. Additionally, understanding and utilizing Kubernetes' autoscaling features ensures that applications can handle varying loads efficiently.

Security within Kubernetes environments cannot be overstated. Best practices here include regularly scanning container images for vulnerabilities, implementing network policies to restrict traffic between pods, and using RBAC to limit permissions to the least privilege necessary. These measures significantly reduce the surface area for potential attacks.

Monitoring and logging are indispensable for maintaining operational excellence. By collecting and analyzing metrics and logs, teams can gain insights into application performance and system health, enabling proactive management of potential issues. Tools that provide alerting based on specific thresholds or patterns can help teams respond quickly to incidents, minimizing impact on users.

Building a culture of learning and collaboration within and beyond the organization plays a vital role in achieving operational excellence. Engaging with the broader Kubernetes community, participating in forums, and attending conferences can provide valuable insights into emerging trends and solutions. Internally, encouraging team members to share knowledge and experiences promotes a continuous improvement mindset.

Operational excellence in Kubernetes is achieved through a combination of automation, efficient resource management, rigorous security practices, diligent monitoring, and a commitment to continuous learning and collaboration. By focusing on these best practices, teams can create Kubernetes environments that are not only resilient and efficient but also poised to evolve with the ever-changing landscape of cloud-native technologies.

Here's a table that organizes performance metrics and benchmarks in a clear and structured format, suitable for evaluating the effectiveness of implemented best practices in Kubernetes environments:

| Metric Type | Metric | Benchmark |
| --- | --- | --- |
| Deployment Frequency | Number of deployments per day/week/month | Increase in frequency without compromising stability |
| Change Lead Time | Time from commit to production | Reduction in lead time over successive iterations |
| **Mean Time to Recovery (MTTR)** | Average time to recover from a failure | Consistent reduction in recovery time |
| Error Rate | % of deployments causing failures | Lower error rates over time |

| Metric Type | Metric | Benchmark |
|---|---|---|
| Resource Utilization Efficiency | CPU and memory usage against allocated resources | High utilization without resource exhaustion |
| Availability/Uptime | % of operational time | Achieving/exceeding industry standards ($\geq 99.9\%$) |
| Security Incident Frequency | Number of security breaches or vulnerabilities | Fewer incidents over time |
| Response Time | Time to respond to system alerts or incidents | Faster response times as processes mature |

Table 9.1 – Performance Metrics and Benchmarks

## Strategic thinking for future-proofing deployments

In the fast-evolving landscape of Kubernetes, planning for the future is as critical as present management. It goes beyond tracking the latest trends; it demands a strategic mindset to ensure deployments are robust, flexible, and poised to capitalize on new opportunities. The essence of future-proofing Kubernetes deployments lies in building systems that can evolve over time, withstand changes in technology, and continue to meet the needs of the business and its customers.

One fundamental aspect of this strategic thinking involves designing deployments with flexibility in mind. This means adopting practices and architectures that allow for easy updates and modifications without significant downtime or rework. For example, using microservices architectures can enable teams to update individual components of an application independently, reducing the risk associated with changes and allowing for more rapid iteration.

Another key strategy is to invest in automation wherever possible. Automation can significantly reduce the manual effort required to manage deployments, from scaling up resources in response to demand to deploying new versions of applications. By automating routine tasks, teams can not only reduce the potential for human error but also free up valuable time to focus on more strategic initiatives that drive the business forward.

Staying informed about advancements in the Kubernetes ecosystem and related technologies is also vital for future-proofing deployments. This doesn't mean chasing every new trend but rather evaluating new tools, features, and practices in the context of how they can enhance or optimize current operations. Engaging with the community through forums, conferences, and user groups can provide insights into how other organizations are adapting to changes and leveraging new technologies effectively.

Equally important is the commitment to ongoing education and skill development within teams. As Kubernetes continues to evolve, ensuring that team members have access to training and resources to update their skills is critical. This not only prepares the organization to adopt new technologies and practices more readily but also helps attract and retain talent by demonstrating a commitment to professional growth.

Embracing a culture of experimentation and feedback allows teams to test new ideas in a controlled manner, learn from the outcomes, and continuously improve their deployments. This might involve piloting new technologies on a small scale before broader adoption or implementing canary deployments to gauge the impact of changes on performance and user experience.

Strategic thinking for future-proofing Kubernetes deployments is about more than just technology; it's about creating an environment where change is anticipated, planned for, and executed effectively. By focusing on flexibility, automation, continuous learning, community engagement, and a culture of experimentation, organizations can ensure their Kubernetes deployments remain robust and relevant, no matter what the future holds.

## Architectural decisions and implications

Making the right architectural decisions is critical when working with Kubernetes, as these choices have long-lasting implications on the system's performance, scalability, and maintainability. The architecture of a Kubernetes deployment acts as its backbone, influencing how well it can meet current needs while also adapting to future demands. Therefore, understanding the potential impacts of these decisions is essential for anyone involved in designing and managing Kubernetes environments.

One of the first considerations should be how to structure applications to take full advantage of Kubernetes' capabilities. Deciding between a monolithic architecture and a microservices architecture, for example, affects not just the development process but also the deployment, scaling, and updating of applications. While microservices offer more flexibility and can improve scalability, they also introduce complexity in terms of networking and data consistency.

Another architectural decision involves selecting the right storage solutions that align with the applications' requirements. Kubernetes offers various storage options, from ephemeral storage for temporary data to persistent volumes that support storage outside the lifecycle of individual pods. The choice among these options should consider factors such as data persistence, performance requirements, and the need for data to be shared among multiple pods.

Networking within Kubernetes is another area where architectural decisions play a critical role. Configuring network policies, choosing load balancers, and deciding on ingress controllers affect how traffic is routed to and from applications, how services within the cluster communicate, and how secure the overall network is. These decisions directly impact the application's accessibility, performance, and security posture.

Considering how to manage state in stateful applications is crucial. Stateful sets and operators in Kubernetes offer mechanisms to manage stateful workloads, ensuring that they maintain a consistent state across restarts and redeployments. However, they also require careful planning around backup, recovery, and scaling strategies to ensure data integrity and availability.

Planning for **disaster recovery (DR)** and **high availability (HA)** is essential. Architectural decisions here involve configuring replication across multiple nodes or even across clusters and geographical regions to ensure that applications remain available and data is not lost in the event of a failure. These strategies must balance the need for availability with the complexity and cost of the chosen solutions.

Remember—architectural decisions made when designing Kubernetes deployments have significant and far-reaching implications. These decisions influence not only the technical aspects of deployment and management but also the ability to respond to changing requirements and challenges over time. Thoughtful consideration of these factors, guided by best practices and an understanding of the specific needs of the applications and the business, is essential for creating robust, scalable, and maintainable Kubernetes environments.

# Applying knowledge to create stable Kubernetes environments

This section explores applying core concepts to establish resilient Kubernetes environments, enhancing security, simplifying architecture, adapting to workload changes, and leveraging tools for optimization and automation.

## Designing for resilience and stability

Ensuring resilience and stability in the Kubernetes ecosystem, where applications and services undergo continuous deployment and updates, is paramount. This involves crafting systems capable of withstanding failures and unexpected issues while minimizing their impact on performance or user experience. The foundation of such systems lies in thoughtful design choices that anticipate potential points of failure and implement safeguards accordingly.

A key strategy in designing for resilience involves implementing redundancy across different levels of the Kubernetes architecture. This means deploying multiple instances of critical components and services, ensuring that if one instance fails, others can take over without disrupting the system's overall functionality. Similarly, distributing these instances across multiple nodes and, if possible, geographic locations can protect against a wider range of failures, from hardware malfunctions to entire data center outages.

Load balancing plays a crucial role in this context, distributing incoming traffic across multiple instances of an application to prevent any single instance from becoming a bottleneck. Kubernetes' built-in load-balancing mechanisms, combined with external load balancers when necessary, can help achieve this balance, ensuring smooth performance even under heavy loads or during an instance failure.

Another aspect of designing for resilience is effective resource management. This involves carefully configuring resource requests and limits for pods to prevent any one application from monopolizing resources, which could lead to system instability. Kubernetes' **Horizontal Pod Autoscaler (HPA)** can automatically adjust the number of pod instances based on current demand, contributing to both stability and efficient resource use.

Properly handling stateful applications in Kubernetes also requires special attention. StatefulSets provide a framework for deploying and managing stateful applications, offering features such as stable, persistent storage and ordered, graceful deployment and scaling. By using StatefulSets and **persistent volume claims** (**PVCs**), developers can ensure that stateful applications maintain their state across restarts or migrations, crucial for applications such as databases that require consistent data.

Monitoring and proactive issue detection are also essential components of a resilient Kubernetes environment. By continuously monitoring application performance and system health, teams can identify and address issues before they escalate into serious problems. Kubernetes offers various monitoring tools and integrates well with external monitoring solutions, allowing teams to set up comprehensive monitoring that covers everything from individual pod health to overall system performance.

In essence, designing Kubernetes environments for resilience and stability requires a multifaceted approach that addresses potential points of failure at multiple levels. By leveraging Kubernetes' features for redundancy, load balancing, resource management, stateful application support, and monitoring, teams can create systems that are robust against failures and capable of maintaining stable operations through various challenges. This ensures that applications remain available and performant, providing a seamless experience for users and a reliable platform for businesses.

## Enhanced security posture and compliance

Ensuring that Kubernetes environments are not only stable but also secure and compliant with relevant regulations is a crucial aspect that demands attention from the outset. The journey to enhancing the security posture within these environments begins with a deep dive into understanding integral components of Kubernetes security mechanisms. This includes setting up RBAC to manage who can access what resources, defining network policies to control the flow of traffic between pods, and ensuring secure communication channels between cluster components.

One of the foundational steps involves the careful management of secrets, such as API keys and passwords, ensuring they are stored securely and accessed safely by applications when needed. Kubernetes offers secrets management capabilities, but leveraging these effectively requires careful planning and implementation to avoid accidental exposure of sensitive information.

Another vital component is adherence to PoLP. This principle dictates that entities, whether users or applications, should only have access to resources necessary for their function, no more. Implementing this within Kubernetes not only minimizes the potential impact of a breach but also aligns with compliance requirements that often mandate strict access controls.

Regularly scanning container images for vulnerabilities before they are deployed is an essential practice. This proactive approach to security helps identify potential security issues early in the development cycle, reducing the risk of deploying vulnerable applications into production environments.

Moreover, ensuring that all communication within the Kubernetes cluster is encrypted is fundamental to safeguarding data in transit. This includes not just data moving between applications and users but also internal communications between Kubernetes components. Encryption helps protect against interception and unauthorized access to sensitive data.

Keeping the Kubernetes environment updated is another critical practice. With new vulnerabilities discovered frequently, ensuring that the Kubernetes version and all applications running on it are up to date is essential for maintaining a strong security posture. This includes applying patches promptly and upgrading to newer versions that offer enhanced security features and fixes for known vulnerabilities.

While implementing these security measures, it is equally important to maintain documentation and evidence of compliance with relevant standards and regulations. This not only assists in demonstrating compliance during audits but also helps in establishing a culture of security awareness and responsibility across the organization.

In practice, enhancing the security posture and ensuring compliance in Kubernetes environments is an ongoing process that involves regular review and adjustment of policies and practices in response to evolving threats and changing regulatory requirements. It requires a balanced approach that incorporates technical solutions, organizational policies, and a continuous commitment to security and compliance excellence.

## Simplification and modularization techniques

When it comes to creating stable Kubernetes environments, the approach of simplification and modularization plays a crucial role. This strategy revolves around breaking down complex systems into smaller, manageable pieces, making them easier to understand, develop, and maintain. In Kubernetes, this can translate into organizing applications into microservices rather than a single, monolithic structure.

Breaking applications into microservices allows teams to update or troubleshoot specific parts of an application without impacting the entire system. This modular approach not only enhances stability by isolating potential problems but also facilitates faster deployment cycles, as smaller changes can be rolled out more quickly and safely.

In addition to application architecture, simplification and modularization also apply to the way resources and configurations are managed in Kubernetes. Using Helm charts, for example, can streamline the deployment of applications by bundling all necessary resources and configurations into a single package that can be managed as one unit. This not only simplifies the deployment process but also ensures consistency across different environments, reducing the potential for errors.

Labels and annotations in Kubernetes serve as another tool for simplification. By tagging resources with labels, operators can organize and manage them more efficiently, applying operations to groups of resources simultaneously. This can greatly reduce the complexity of managing large numbers of resources, making the environment easier to oversee and control.

Furthermore, adopting a GitOps approach, where infrastructure and application configurations are stored in **version control systems** (**VCSs**), enables teams to manage their Kubernetes environments using the same tools and practices they use for **source code management** (**SCM**). This not only simplifies the management process but also enhances transparency and auditability, as changes are tracked and can be reviewed through pull requests.

It's also essential to leverage Kubernetes' own features for modularization, such as namespaces, to segment resources within the same cluster. This allows for the logical separation of environments, applications, or teams within a single Kubernetes cluster, simplifying management and enhancing security by limiting the scope of resources and permissions.

Implementing these simplification and modularization techniques requires careful planning and consideration of the specific needs and contexts of each application and team. However, by making these principles a core part of the approach to Kubernetes deployment and management, teams can create more manageable, scalable, and stable environments that are easier to develop, maintain, and scale over time.

## Adaptive strategies for evolving workloads

In the dynamic world of Kubernetes, workloads are constantly evolving, driven by changing user demands, technological advancements, and the need for businesses to stay competitive. To keep pace with these changes, adopting adaptive strategies that allow for the seamless evolution of workloads is essential. This involves setting up environments that can quickly respond to new requirements without requiring extensive reconfiguration or downtime.

One key approach to achieving this flexibility is through the use of autoscaling. Kubernetes provides native autoscaling features, such as HPA and **Vertical Pod Autoscaler** (**VPA**), which automatically adjust the number of pods or their resource limits based on observed metrics such as CPU usage or memory consumption. By leveraging these tools, applications can maintain optimal performance levels even as workload demands fluctuate.

Another strategy involves the implementation of rolling updates and canary deployments. Rolling updates allow for new versions of applications to be gradually rolled out without interrupting the service, ensuring that any potential issues affect only a small portion of users and can be quickly addressed. Canary deployments take this one step further by routing a small amount of traffic to the new version for testing before it's fully deployed, minimizing the risk associated with changes.

Container orchestration environments thrive on the principle of immutability, where changes are made by replacing containers rather than modifying them directly. This approach simplifies updates and rollbacks, as new container images can be deployed and scaled up while the old ones are scaled down, ensuring that the system can adapt quickly to new requirements without the risk of state corruption or configuration drift.

Furthermore, leveraging cloud-native storage solutions that offer dynamic provisioning can significantly enhance the adaptability of workloads. Such solutions automatically provision storage as needed by applications, ensuring that storage requirements can scale with the application without manual intervention.

To effectively implement these adaptive strategies, it's also crucial to have a robust monitoring and alerting system in place. Monitoring provides visibility into the performance and health of applications and infrastructure, enabling teams to proactively adjust resources and configurations in response to observed metrics and trends. Alerting ensures that any potential issues are quickly identified and addressed, maintaining the stability and reliability of the environment.

Embracing adaptive strategies for evolving workloads requires a proactive mindset and a willingness to embrace new tools and practices. By building Kubernetes environments with flexibility and adaptability at their core, organizations can ensure that their applications remain resilient, performant, and aligned with business objectives, regardless of how workloads change over time.

## Tooling for optimization and automation

The role of tooling in managing Kubernetes environments cannot be overstated. Both optimization and automation stand at the forefront of effective Kubernetes management, allowing teams to streamline operations, reduce manual effort, and significantly improve the reliability and efficiency of their deployments. The choice of tools in a Kubernetes ecosystem plays a pivotal role in achieving these goals.

Optimization tools are designed to analyze the performance and resource utilization of applications running in Kubernetes, identifying opportunities to improve efficiency. These might include solutions that offer insights into pod resource usage, network throughput, or storage performance. By leveraging such tools, teams can pinpoint bottlenecks or over-provisioned resources, adjusting configurations to better match the actual needs of the applications, thereby reducing waste and improving overall performance.

On the other hand, automation tools focus on reducing the manual overhead associated with deploying, managing, and scaling applications in Kubernetes. This includes CI/CD pipelines that automate the process of building, testing, and deploying applications. Automation also extends to scaling, with tools that automatically adjust the number of pods based on traffic patterns, and to self-healing mechanisms that automatically replace failed pods or nodes to ensure HA.

Another important category of tooling encompasses security and compliance. These tools scan container images for vulnerabilities, enforce security policies at runtime, and ensure that deployments comply with industry standards and regulations. By automating security checks and compliance monitoring, organizations can maintain a strong security posture without adding significant manual effort.

Monitoring and logging tools are also crucial, providing visibility into the health and performance of applications and the underlying infrastructure. These tools collect metrics and logs, presenting them through dashboards or alerting administrators to potential issues before they impact users. Effective monitoring and logging are essential for the proactive management of Kubernetes environments, allowing teams to quickly respond to changes in application behavior or performance.

Selecting the right set of tools requires a careful evaluation of the specific needs of the organization and its applications. It often involves integrating multiple tools into a cohesive toolchain that covers the entire life cycle of applications, from development and deployment to operation and optimization.

The integration of these tools into the Kubernetes environment should be done with an eye toward flexibility and scalability, ensuring that they can adapt to the evolving needs of the organization and the dynamic nature of Kubernetes workloads.

By focusing on tooling for optimization and automation, organizations can create Kubernetes environments that are not only more manageable and efficient but also more resilient and responsive to the needs of the business.

# Encouraging a culture of continuous improvement

This section concentrates on building a culture of continuous improvement in Kubernetes management, highlighting the development of a learning mindset, proactive practices, feedback mechanisms, innovation, and the significant impact of leadership on strategy.

## Cultivating a learning and improvement mindset

In a landscape where Kubernetes holds a pivotal position in application deployment and management, the rapid pace of technological advancement underscores the need for continuous learning and enhancement. Within this domain, cultivating a mindset oriented toward ongoing learning isn't merely advantageous—it's indispensable for remaining relevant and efficient. This ethos of perpetual development guarantees that as Kubernetes progresses, the skills and approaches of its practitioners evolve in tandem.

Starting with individual team members, the encouragement to actively seek out new information, experiment with emerging technologies, and reflect on the outcomes of these explorations is vital. This might involve dedicating time each week to learning new aspects of Kubernetes, participating in workshops, or contributing to open source projects related to Kubernetes. Such activities not only enhance individual knowledge but also bring fresh ideas and perspectives back to the team.

On a team level, promoting a culture of sharing knowledge plays a crucial role. Regularly scheduled sessions where team members can share insights from recent learnings, discuss challenges faced in current projects, or conduct post-mortem analyses of past deployments help to disseminate knowledge throughout the team. This not only helps in leveling up the team's collective expertise but also fosters a supportive environment where learning from mistakes is valued as much as celebrating successes.

For the organization as a whole, investing in formal training and certification programs for Kubernetes can demonstrate a commitment to professional development. Providing access to resources such as online courses, attending industry conferences, or bringing in external experts for specialized training sessions can equip teams with the knowledge and skills needed to navigate the complexities of Kubernetes effectively.

Embracing tools and practices that support experimentation and learning can further enhance this culture. Implementing sandbox environments where team members can safely experiment with new configurations, architectures, or technologies without the risk of affecting production systems allows for hands-on learning and innovation.

The process of cultivating a learning and improvement mindset within a Kubernetes context is ongoing. It requires deliberate actions from individuals, support and encouragement from team leaders, and strategic investment from the organization. By making learning and continuous improvement a core part of the culture, teams can ensure that they not only keep pace with the rapid developments in Kubernetes but also leverage these advancements to drive better outcomes for their deployments and the business as a whole.

## Proactive practices to anticipate challenges

Creating a culture where teams actively prepare for potential obstacles is crucial in the fast-paced environment of Kubernetes. This means setting up practices and workflows that not only address issues as they arise but also anticipate and mitigate them before they impact operations. By being proactive, organizations can maintain a high level of service reliability and performance, even as they evolve and scale their Kubernetes deployments.

One key approach is implementing comprehensive monitoring and alerting systems. These tools provide real-time insights into the health and performance of applications and infrastructure, enabling teams to detect and address anomalies before they escalate into more significant problems. By carefully defining metrics and thresholds, teams can create a detailed picture of normal operations, making it easier to spot when something deviates from the expected.

Another practice involves conducting regular risk assessments and scenario-planning exercises. By evaluating the Kubernetes environment and applications for potential vulnerabilities or failure points, teams can develop strategies to mitigate these risks. This might include everything from improving security measures to planning for DR, ensuring the organization is prepared for various challenges.

Automation plays a significant role in being proactive. Automating routine tasks, such as deployments, scaling, and backups, not only reduces the potential for human error but also frees up team members to focus on more strategic work. Automation can also extend to self-healing mechanisms where the system automatically responds to certain types of failures, further enhancing the resilience of Kubernetes environments.

Engaging with the broader Kubernetes community is another way to stay ahead of challenges. By sharing experiences and learning from the successes and failures of others, teams can gain insights into potential issues before they encounter them firsthand. This community engagement can take many forms, from participating in forums and attending conferences to contributing to open source projects.

Encouraging open communication and collaboration within teams is essential. By creating an environment where team members feel comfortable sharing their observations, concerns, and ideas, organizations can tap into a wealth of knowledge and perspectives. This collective approach to problem-solving not only helps in anticipating challenges but also fosters a sense of ownership and accountability among team members.

By implementing these proactive practices, organizations can create Kubernetes environments that are not only more stable and secure but also adaptable to the changing needs of the business. This proactive mindset ensures that teams are always prepared for the future, ready to tackle challenges head-on and continue delivering value without interruption.

## Effective feedback mechanisms

Establishing mechanisms for collecting and acting on feedback is vital for continuous improvement. This involves creating channels through which team members can share their observations, experiences, and suggestions regarding the Kubernetes environment and its workflows. By making it easy for everyone involved to offer feedback, organizations can identify areas for enhancement, innovate more effectively, and resolve issues more swiftly.

One approach is to implement regular review sessions where teams discuss the performance of deployments, share challenges faced during development and operation phases, and suggest improvements. These sessions can be structured around specific projects or be more open-ended to cover a broader range of topics. The key is to ensure that these discussions are inclusive, encouraging participation from all team members, regardless of their role or level of experience.

Another valuable feedback mechanism is the use of issue-tracking and project management tools. These platforms allow team members to report problems, suggest enhancements, and track the progress of their implementation. By maintaining transparency in the process, everyone can see which suggestions are being acted upon, fostering a sense of ownership and accountability.

Surveys and feedback forms distributed after completing significant milestones or projects can also provide insights into areas that might not be covered in daily communications or meetings. These tools can gather anonymous feedback, offering a safe space for more candid responses that might not be shared openly. Analyzing data from these surveys can highlight patterns and opportunities for improvement that might not be immediately obvious.

Incorporating feedback into the development life cycle through CI/CD pipelines is another strategy. Automated testing, performance benchmarks, and **user acceptance testing** (**UAT**) phases can all serve as feedback mechanisms, providing quantitative data on the impact of changes. By closely integrating feedback into the development process, teams can more rapidly iterate on and refine their applications and services.

Creating a knowledge base where lessons learned, best practices, and feedback outcomes are documented and shared can serve as a long-term resource for the team. This repository of knowledge not only helps in onboarding new members but also serves as a reference point for planning future projects.

Effective feedback mechanisms are essential for adapting to fast-paced changes inherent in Kubernetes environments. They enable teams to learn from their experiences, continuously improve their processes, and maintain a high level of performance and reliability in their deployments. By prioritizing communication and feedback, organizations can cultivate a culture of continuous improvement where every team member feels valued and empowered to contribute to the success of their Kubernetes initiatives.

## Encouraging innovation and experimentation

Creating an environment where innovation and experimentation are not just allowed but actively encouraged is crucial for teams working with Kubernetes. This approach helps to ensure that new ideas and technologies can be explored, potentially leading to more efficient, resilient, and effective Kubernetes deployments. The nature of Kubernetes, with its flexibility and extensive ecosystem, makes it an ideal platform for testing new concepts and approaches.

One way to encourage this atmosphere is by setting aside dedicated time and resources for team members to work on projects or ideas that interest them, even if these are not directly related to their day-to-day tasks. These projects can provide valuable learning opportunities and may uncover innovative solutions to existing problems or identify new ways to use Kubernetes more effectively.

Another strategy is to create a safe space for failure. Understanding that not every experiment will be successful but that each attempt provides a learning opportunity is key. By removing the stigma associated with failure, team members are more likely to take risks and try out new ideas. This can lead to breakthroughs that significantly improve the efficiency and reliability of Kubernetes environments.

Implementing mechanisms to share outcomes from these experiments, whether successful or not, is also important. This could take the form of regular show-and-tell sessions, where team members present their projects and findings. These sessions not only spread knowledge and spark further ideas but also celebrate the effort and creativity involved in innovation.

Engaging with the wider Kubernetes community can inspire innovation and experimentation. Participating in forums, contributing to open source projects, or attending conferences can expose team members to new perspectives and ideas that can be adapted and applied to their own work.

By encouraging a culture of innovation and experimentation, organizations can ensure that their Kubernetes environments continue to evolve and improve. This not only leads to more efficient and effective deployments but also contributes to a more engaged and motivated team.

## Leadership's role in Kubernetes strategy

The success of Kubernetes initiatives within any organization significantly hinges on the role played by its leaders. These individuals are not just decision-makers but also visionaries who guide the strategic direction of Kubernetes deployments. Their involvement can make a profound difference in how these technologies are adopted and utilized across teams and projects.

Leaders are tasked with setting clear goals and expectations for Kubernetes initiatives. By defining what success looks like, they provide teams with a direction and purpose, aligning Kubernetes projects with broader business objectives. This clarity helps in prioritizing efforts and resources effectively, ensuring that work done delivers real value to the organization.

Moreover, leaders have the responsibility to ensure that their teams have the necessary resources and support to succeed. This includes providing access to training and learning opportunities to keep skills up to date with the latest Kubernetes developments. It also involves investing in the right tools and technologies that enable teams to implement, manage, and scale Kubernetes environments efficiently.

Creating an open and inclusive culture is another critical aspect of leadership. By encouraging open communication, leaders can foster an environment where feedback is valued and different perspectives are welcomed. This openness not only helps in identifying and addressing challenges quickly but also in capturing diverse ideas that can lead to innovative solutions.

Leaders also play a crucial role in promoting collaboration both within and outside the organization. By breaking down silos and encouraging cross-functional teams to work together on Kubernetes projects, leaders can leverage the full range of skills and expertise available. Additionally, engaging with the external Kubernetes community allows leaders to bring in external insights and best practices, further enriching the organization's knowledge base.

Leaders must lead by example, demonstrating a commitment to continuous improvement and innovation. By actively participating in Kubernetes initiatives, staying informed about new developments, and sharing their own learnings, leaders can inspire their teams to strive for excellence.

In essence, the role of leadership in shaping Kubernetes strategy is multifaceted, involving goal setting, resource allocation, culture building, collaboration, and personal involvement. Through their actions and decisions, leaders have the power to drive the successful adoption and optimization of Kubernetes, enabling their organizations to fully realize the benefits of this transformative technology.

## Summary

This chapter brought together crucial insights from the entire book, outlining the identification of Kubernetes anti-patterns, addressing key challenges with effective solutions, and emphasizing best practices for operational excellence. It detailed strategies for future-proofing deployments and the impact of architectural decisions. The discussion extended to practical applications for creating stable environments, including resilience, security enhancements, and the use of automation tools. Finally, it underscored the importance of cultivating an environment conducive to continuous improvement, innovation, and the pivotal role of leadership in guiding Kubernetes strategy.

# Index

packtpub.com

Subscribe to our online digital library for full access to over 7,000 books and videos, as well as industry leading tools to help you plan your personal development and advance your career. For more information, please visit our website.

## Why subscribe?

- Spend less time learning and more time coding with practical eBooks and Videos from over 4,000 industry professionals

- Improve your learning with Skill Plans built especially for you

- Get a free eBook or video every month

- Fully searchable for easy access to vital information

- Copy and paste, print, and bookmark content

Did you know that Packt offers eBook versions of every book published, with PDF and ePub files available? You can upgrade to the eBook version at packtpub.com and as a print book customer, you are entitled to a discount on the eBook copy. Get in touch with us at customercare@packtpub.com for more details.

At www.packtpub.com, you can also read a collection of free technical articles, sign up for a range of free newsletters, and receive exclusive discounts and offers on Packt books and eBooks.

# Other Books You May Enjoy

If you enjoyed this book, you may be interested in these other books by Packt:

**Mastering Kubernetes**

Gigi Sayfan

ISBN: 978-1-80461-139-5

- Learn how to govern Kubernetes using policy engines
- Learn what it takes to run Kubernetes in production and at scale
- Build and run stateful applications and complex microservices
- Master Kubernetes networking with services, Ingress objects, load balancers, and service meshes
- Achieve high availability for your Kubernetes clusters
- Improve Kubernetes observability with tools such as Prometheus, Grafana, and Jaeger
- Extend Kubernetes with the Kubernetes API, plugins, and webhooks

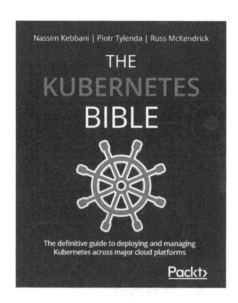

**The Kubernetes Bible**

Nassim Kebbani, Piotr Tylenda, Russ McKendrick

ISBN: 978-1-83882-769-4

- Manage containerized applications with Kubernetes
- Understand Kubernetes architecture and the responsibilities of each component
- Set up Kubernetes on Amazon Elastic Kubernetes Service, Google Kubernetes Engine, and Microsoft Azure Kubernetes Service
- Deploy cloud applications such as Prometheus and Elasticsearch using Helm charts
- Discover advanced techniques for Pod scheduling and auto-scaling the cluster
- Understand possible approaches to traffic routing in Kubernetes

## Packt is searching for authors like you

If you're interested in becoming an author for Packt, please visit `authors.packtpub.com` and apply today. We have worked with thousands of developers and tech professionals, just like you, to help them share their insight with the global tech community. You can make a general application, apply for a specific hot topic that we are recruiting an author for, or submit your own idea.

## Share Your Thoughts

Now you've finished *Kubernetes Anti-Patterns*, we'd love to hear your thoughts! Scan the QR code below to go straight to the Amazon review page for this book and share your feedback or leave a review on the site that you purchased it from.

`https://packt.link/r/1835460682`

Your review is important to us and the tech community and will help us make sure we're delivering excellent quality content.

# Download a free PDF copy of this book

Thanks for purchasing this book!

Do you like to read on the go but are unable to carry your print books everywhere?

Is your eBook purchase not compatible with the device of your choice?

Don't worry, now with every Packt book you get a DRM-free PDF version of that book at no cost.

Read anywhere, any place, on any device. Search, copy, and paste code from your favorite technical books directly into your application.

The perks don't stop there, you can get exclusive access to discounts, newsletters, and great free content in your inbox daily

Follow these simple steps to get the benefits:

1.  Scan the QR code or visit the link below

https://packt.link/free-ebook/9781835460689

2.  Submit your proof of purchase
3.  That's it! We'll send your free PDF and other benefits to your email directly

www.ingramcontent.com/pod-product-compliance
Lightning Source LLC
Chambersburg PA
CBHW080635060326
40690CB00021B/4946